FM 23-100

WAR DEPARTMENT FIELD MANUAL

TANK AND TANK DESTROYER GUNNERY FIELD MANUAL

RESTRICTED. DISSEMINATION OF RESTRICTED MATTER. No person is entitled solely by virtue of his grade or position to knowledge or possession of classified matter. Such matter is entrusted only to those individuals whose official duties require such knowledge or possession. (See also paragraph 23b, AR 380-5, 15 March 1944.)

BY WAR DEPARTMENT • JANUARY 1946

ISBN#978-1-937684-56-3
www.PeriscopeFilm.com

DISCLAIMER:

This manual is sold for historic research purposes only, as an entertainment. It contains obsolete information and is not intended to be used as part of an actual operation or maintenance training program. No book can substitute for proper training by an authorized instructor.

ISBN#978-1-937684-56-3
www.PeriscopeFilm.com

WAR DEPARTMENT FIELD MANUAL

FM 23-100

This manual supersedes FM 17-12, Tank Gunnery, 10 July 1944.

TANK AND TANK DESTROYER GUNNERY

WAR DEPARTMENT • JANUARY 1946

RESTRICTED. *DISSEMINATION OF RESTRICTED MATTER.* No person is entitled solely by virtue of his grade or position to knowledge or possession of classified matter. Such matter is entrusted only to those individuals whose official duties require such knowledge or possession. (See also paragraph 23b, AR 380-5, 15 March 1944.)

United States Government Printing Office
Washington : 1945

WAR DEPARTMENT
WASHINGTON 25, D. C., 7 January 1946

FM 23–100, Tank and Tank Destroyer Gunnery, is published for the information and guidance of all concerned.

[AG 300.7 (13 Nov 45)]

BY ORDER OF THE SECRETARY OF WAR:

DWIGHT D. EISENHOWER
Chief of Staff

OFFICIAL:

EDWARD F. WITSELL
Major General
Acting The Adjutant General

DISTRIBUTION:

AAF (2); AGF (40); ASF (2); T of Opns (5); Arm & Sv Bd (1); Sp Sv Sch 2, 6 (50), 7, 17 (250); USMA (5); Tng C 17 (250); A (5); CHQ (5); D (5); B 2 (3); R 2, 6, 7, 17, 18 (5); Bn 17, 18 (5); C 17 (10), except T/O & E: 17–60–1 (0). T/O & E: 2–10–1 (5); 2–26 (5); 2–27 (5); 18–28 (5); 2–28 (10); 7–14 (10); 7–19 (10); 18–27 (10).

Refer to FM 21–6 for explanation of distribution formula.

CONTENTS

PART ONE. GENERAL.

PART TWO. TECHNIQUE OF GUNNERY.

RESTRICTED

This manual supersedes FM 17-12, Tank Gunnery, 10 July 1944.

PART ONE

GENERAL

CHAPTER 1

INTRODUCTION

1. PURPOSE AND SCOPE. This manual is a guide for gunnery training for tank and tank destroyer crews. Parts One and Two describe the direct fire gunnery technique for cannon laid by a one-man sighting system. Parts Three and Four outline the technique of employing direct fire cannon in indirect fire and as reinforcing artillery. Part Five describes the gunnery technique to be employed with auxiliary weapons.

2. DEFINITIONS. Following are the meaning of several terms as used in this manual:

a. Crew commander. Commander of a tank or tank destroyer.

b. Gun motor carriage. Any carriage for a self-propelled gun, including tanks, tank destroyers, and armored cars.

c. Medium velocity cannon. Cannon having a muzzle velocity between 1,250 and 2,500 feet per second.

Note. For terms not defined in this manual, see TM 20-205; for list of training publications, see FM 21-6; for training films, film strips, and film bulletins, see FM 21-7; and for training aids, see FM 21-8.

d. High velocity cannon. Cannon having a muzzle velocity in excess of 2,500 feet per second.

3. DUTIES OF GUNNERY OFFICERS. Commanders of all echelons are responsible for the gunnery training of their respective units in accordance with current unit training programs. Their gunnery duties are to—

a. Supervise and coordinate the gunnery training within the unit.

b. Require proper methods of sight adjustment.

c. Insure that weapons receive proper care and maintenance.

d. Supervise the training of officers and noncommissioned officers in the preparation and conduct of fire and in the coordination of fires within the unit.

e. Insure the following in training:

(1) *Thoroughness.* Training should never leave mistakes to be discovered on the battlefield.

(2) *Progressive training.* Gunners should be permitted to proceed to a second exercise only when the first has been mastered.

(3) *Periodic review.* Every step of training must be reviewed before, during, and after movement to the next phase.

(4) *Discipline.* Strict discipline must be maintained so that even under the stress of battle the well disciplined gunner will habitually apply the practices of good gunnery.

(5) *Teamwork.* Instruction should be conducted for crews, sections, platoons — each working under its *regularly assigned leader.* Ef-

fective gunnery is built on efficient gun crews.

(6) *Explanation of battle application.* The reasons for every step of training and its battle value must be explained thoroughly.

f. Supervise all firing.

4. IMPORTANCE OF FIRE CONTROL. The company commander controls expenditure of ammunition as closely as possible. The platoon leader exercises even stricter control. Except in the closest and most unexpected engagements, he should personally supervise the firing of the whole platoon. Where practicable, the crew commander selects every target personally and carefully controls the firing of each round. *All crew members should be alert at all times to designate targets to the crew commander.*

5. CRITIQUE OF GUNNERY PROBLEM OR FIRING EXERCISE. A critique must follow each problem or exercise. A good critique is clear, concise, and offers constructive criticism of errors. To be of value, it must immediately follow the exercise and should cover the following items:

a. A statement of the mission, including a tactical description of the target; for example: "The mission was to neutralize a section of machine guns which was holding up the advance of the supported unit."

b. A statement of the proper method of attack, including the selection of ammunition.

c. A statement as to whether the mission was accomplished and whether in a satisfactory or unsatisfactory manner.

d. A statement of the principal points, good and bad, illustrated by the problem.

e. An explanation of the problem's application to the battlefield. The gravity of carrying training errors into combat must be fully explained.

f. An opportunity for questions by all who took part in the problem or observed it.

6. OUTLINE OF BASIC TRAINING. **a.** During basic training, each crew member should be trained equally in all crew positions. At the conclusion of basic training, permanent crews should be assigned to combat vehicles and trained as teams.

b. When each man has been assigned a permanent position, effort should be concentrated on his achieving perfection in that position. Further training in the duties of other positions should be conducted only when each man has become proficient at his own job.

c. Care will be exercised in the selection of appropriate training areas.

d. The vehicles will be fully stowed, including individual equipment, during all advanced training. If all the stowage items are not available, substitutes should be constructed.

e. The crew should be accustomed to combat conditions by realistic training. Combat practice firing problems should be used. (See FM 17–15.)

CHAPTER 2

RANGE DETERMINATION

Section I. GENERAL

7. DETERMINING RANGE. The effectiveness of the opening shot and the number of rounds required to engage a target effectively depend upon the accuracy with which the range is determined. This paragraph lists several methods of determining range. They are listed in order by the degree of accuracy obtained, without regard to the time required or tactical feasibility. The most accurate method that the situation will permit should be the one used in the field.

a. Intersection or short base method. Range may be determined by intersection, using an angle at each end of a measured base line and the application of the mil relation or short base method. A detailed discussion of this technique is contained in section IV, this chapter.

b. Use of range finder. The range finder is used as prescribed in pertinent technical manuals.

c. Registration. Range for any particular weapon may be determined by firing at the target or any object close to the target. The range selected is the range at which the projectile strikes the target or close object. This procedure may be used when the fire of the platoon is to be concentrated on the target or in preparing a range card for a stabilized position.

d. Firing coaxial machine gun. For vehicles so equipped, range may be determined by firing the coaxial machine gun. The machine gun is fired with an estimated range and the fire is adjusted until a hit is obtained. The point on the reticle at which the strike appears is the range setting for the machine gun. The gunner refers to the aiming data chart (figs. 18 and 19) and determines the corresponding range setting for the cannon. Unless the strike can be observed, the range at which this method can be used is limited by the tracer burn-out point.

e. Use of binoculars or sight. If the width or height of an object is known, the range to it may be determined in yards by the use of field glasses and the application of the mil formula.

f. Estimation by eye. This is the usual means of determining range. It is also the least accurate. A detailed discussion of this method is contained in section II, this chapter.

g. Other methods. These methods are not listed with regard to accuracy but are cited merely as possible methods of determining range.

(1) From a map or photomap, when the gun and target can be located.

(2) By information from friendly troops.

Section II. ESTIMATION BY EYE

8. INSTRUCTION. Accuracy in range estimation is attained only after frequent practice under a great variety of conditions. The following rules are for instructors conducting training in range estimation (fig. 1):

a. Know the true ranges.

b. Always announce the true range to each target immediately after the men have made their estimates.

c. Move to new terrain as often as possible. If this is not done, the estimates will become comparative judgments against known ranges.

d. Employ typical combat targets. Include trees, bushes, rocks, ridge lines, edges of woods, changes in vegetation, antitank guns, tanks, trucks, houses, buildings, bridges, and road junctions.

e. Vary the backgrounds. Use such backgrounds as the sky, trees, open fields, and bodies of water.

f. Start training in estimation with the naked eye. Proceed to training with sights, periscopes, and field glasses only after the men have become proficient without them.

Figure 1. Instruction in range estimation.

g. Conduct instruction in the estimation of ranges varying from 100–5,000 yards. Devote more time, however, to ranges of less than 2,000 yards.

h. Teach the men to make each estimate an independent one. A distance of 800 yards does not look like four times the distance of 200 yards which lies directly in front of the observer. For example: the first 200 yards appears longer than the 200 yards between 600 and 800 yards. A range of 800 yards should always look like 800 yards.

i. Teach the men to study the ground between themselves and the target.

9. PRACTICE. Practice should be conducted under conditions which will illustrate the following principles:

a. Targets appear nearer and the range is underestimated when—

(1) The object is in a bright light.

(2) The color of the object contrasts sharply with the color of the background.

(3) Looking over water, snow, or a uniform surface such as a wheat field or desert.

(4) Looking down a straight road or along a railroad track.

(5) Looking downward from a height.

(6) Looking across a depression, most of which is hidden.

b. Targets seem more distant and the range is overestimated when—

(1) Looking across a depression, most of which is visible.

(2) There is fog, rain, or the light is poor.

(3) Only part of the target is visible.

(4) Looking from low ground toward higher ground.

10. SUMMARY. Two general principles cover most of the facts outlined in paragraph 9.

a. The first principle relates to the optical illusion illustrated by a broken or intercepted line. (See fig. 2.) The illusion exists because the space between *A* and *B* is broken up by the vertical lines and appears greater. The space between *B* and *C* is unbroken, and the line appears shorter. Applied to terrain, the distance appears greater and the range is overestimated when the ground between the observer and the target is broken by trees, scattered vegetation, rocks, and other objects. The distance appears shorter and the range is underestimated when the ground between the observer and the target is not visible or presents a smooth surface.

b. The second principle concerns the relative visibility of the target. On a clear day, when the observer looks over a landscape which includes very distant objects such as hills, the near objects are clear while those at a much greater distance

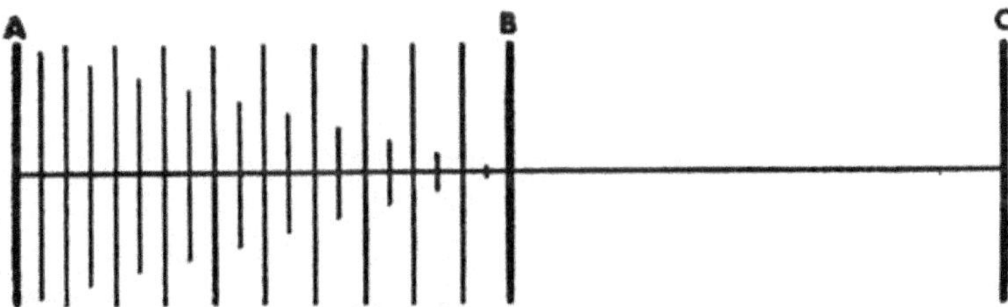

Figure 2. Is line AB longer than line BC?

seem relatively unclear and hazy. This fact has been so generally observed that it unconsciously affects judgment of distance. Distinct objects seem to be close, and ranges to them tend to be underestimated. Hazy objects seem more distant, and ranges to them tend to be overestimated.

c. These two principles illustrate the necessity for estimating ranges under varied conditions and on different types of terrain.

d. The fact that range estimates must be made *spontaneously* will be continuously emphasized. The men should be trained so thoroughly and under such a variety of conditions that compensation for the two principles will enter unconsciously into their range estimates.

Section III. MIL FORMULA

11. GENERAL. **a.** Range may be determined by use of the mil formula.

b. In gunnery, angles are measured in mils. All fire control equipment is graduated in mils. A clear understanding of the mil is necessary before learning the operation of fire control equipment or the use of the mil formula.

12. DESCRIPTION AND USE. **a.** The full circle contains 6,400 mils. Therefore, a mil is 1/6400 part of the circle. That is—

$360° = 6{,}400$ mils.

$1° =$ approximately 18 mils (actually 17.777+).

b. Practically, it is also true that 1 mil is subtended by 1 yard at a distance of 1,000 yards. (See fig. 3.) That is—

1 mil is subtended by 1 yard at 1,000 yards.

1 mil is subtended by 2 yards at 2,000 yards.

Similarly—

150 mils are subtended by 150 yards at 1,000 yards.

150 mils are subtended by 300 yards at 2,000 yards.

These relationships are accurate enough for gunnery calculations for any angle less than 400 mils.

c. Knowing any two of the three elements involved, the third can be computed by the mil formula as shown in figure 4.

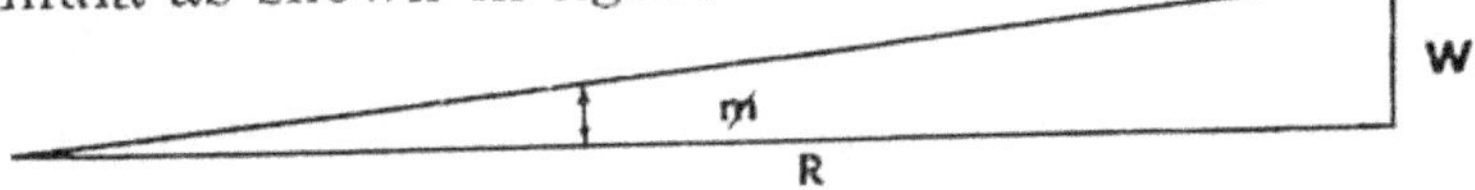

W = WIDTH SUBTENDED IN YARDS
R = RANGE IN THOUSANDS OF YARDS
m̸ = ANGULAR MEASUREMENT IN MILS

Figure 3. Mil formula.

m̸ (MILS) = W (WIDTH) / R (RANGE IN THOUSANDS)

EXAMPLE 1: COMPUTE THE ANGLE

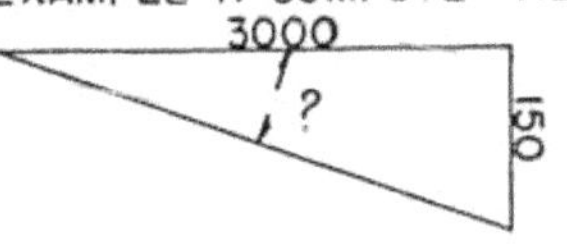

ANS: ANGLE IS 50 MILS
(150 YDS / 3 = 50 m̸)

EXAMPLE 2: COMPUTE THE BASE WIDTH

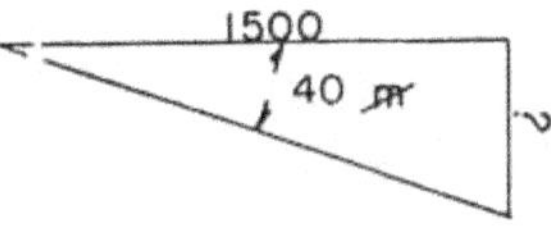

ANS: BASE IS 60 YDS
(40 m̸ = 60 YDS / 1.5)

EXAMPLE 3: COMPUTE THE RANGE

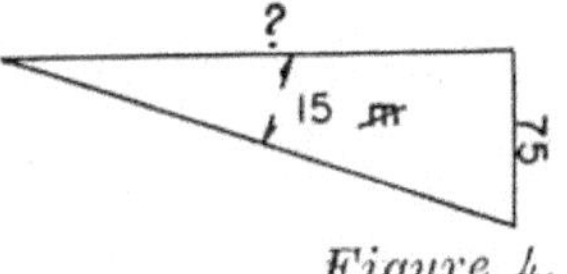

ANS: RANGE IS 5000 YDS
(15 m̸ = 75 YDS / 5)

Figure 4. Using mil formula.

d. A simplification of the mil formula, which permits a much speedier solution, is the *WORM* formula. The letters *W–O–R–M* stand for *W* over *Rm*, or $\frac{W}{Rm}$. The letters *W*, *R*, and *m* stand, respectively, for width in yards, range in thousands of yards, and width in mils. When the values of two of the elements are known, that of the unknown may be rapidly found by substituting these known values for the appropriate letters in the fraction $\frac{W}{Rm}$, covering the letter whose value is unknown and solving for the visible figures. The result is the value of the third, or unknown, element.

e. *Example:* The known length of a tank is 8 yards. It measures 5 mils in the binoculars. What is the range? The known values 8 and 5 are substituted respectively for *W* and *m*. Covering *R* and solving the visible fraction gives $\frac{8}{5}$, or 1.6. Since *R* is always in thousands, the range is 1,600 yards.

Section IV. INTERSECTION

13. GENERAL. Intersection is an effective and accurate means of range determination which employs two angle-measuring instruments, such as azimuth indicators, panoramic sights, or aiming circles.

14. PROCEDURE. **a.** The instruments to be used should have no obstructions between them. When azimuth indicators are used, care must be taken that the gun carriages are approximately level. The distance between the instruments must be at least 20 percent of the range to the target and each approximately the same distance from the target.

b. The length of the base line, *AB* (fig. 5), is measured accurately.

c. Each operator sights his instrument on the other operator's instrument or gun. At night, it is often better to sight on the target first. In either case, scales are initially set at zero. When using the azimuth indicator at night, it may be necessary to place a light in the firing pin well of each gun. When this is done, care must be taken to prevent the enemy from seeing the light.

d. The operators then sight their instruments on the target. The size of the angles *A* and *B* (fig. 5) is determined. Depending upon which instrument is used, this step produces either the desired angles or their supplements. In solving the triangle, extreme care must be taken that the figures used actually are the interior angles *A* and *B*.

e. The triangle may be solved—

(1) By use of the mil formula.

(2) By short-base methods.

f. Solving by the short-base method is far more accurate. Further, it is not subject to the limitation of the mil formula. The procedure is described in FM 6–40.

g. When using the mil formula, the base line should be selected so that the angle *C* is not less than 150 mils nor greater than 400 mils. The

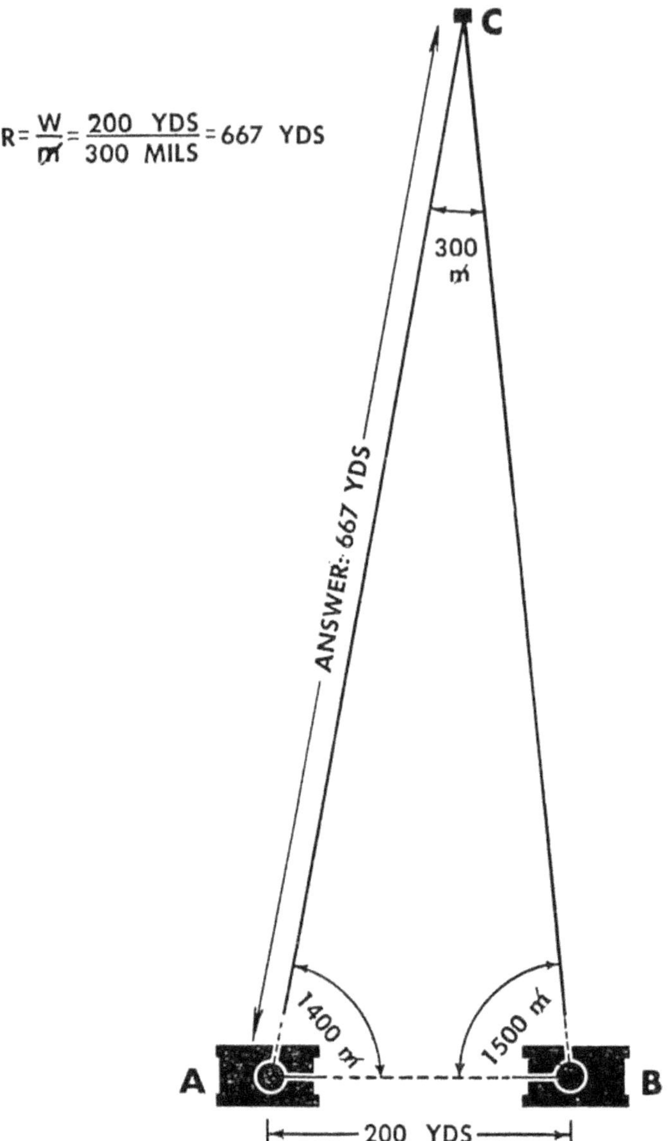

Figure 5. Determining range by intersection.

angles A and B, respectively, should not vary from right angles (1,600 mils) by more than 300 mils. The mil formula is solved as described in section III, this chapter. Values are substituted in the mil formula as follows:

(1) W equals the length of the base line.

(2) m equals the angle C.

(3) C equals 3,200 minus (A plus B).

h. *Example:* The known distance between the tanks at A and B (fig. 5) is 200 yards. Angle A is measured to be 1,400 mils; angle B is measured to be 1,500 mils. By subtracting the sum of angles A and B from 3,200 mils, the value of angle C is determined to be 300 mils. Hence the range from tank A to target C (in thousands of yards) $= R = \frac{W}{m} = \frac{200 \text{ yards}}{300 \text{ mils}} = 667$ yards.

Section V. RANGE CARDS

15. GENERAL. A range card (fig. 6) is a diagrammatic sketch of a given sector showing gun positions, direction of magnetic north, field of fire including dead space, and ranges to prominent terrain features. Crew commanders should prepare range cards for each gun position.

16. PURPOSE. The purpose of a range card is threefold:

a. It enables the crew commander to deliver fire promptly and accurately on likely targets.

b. It aids the platoon leader, through use of duplicate cards, in coordinating the fires of the guns in his platoon.

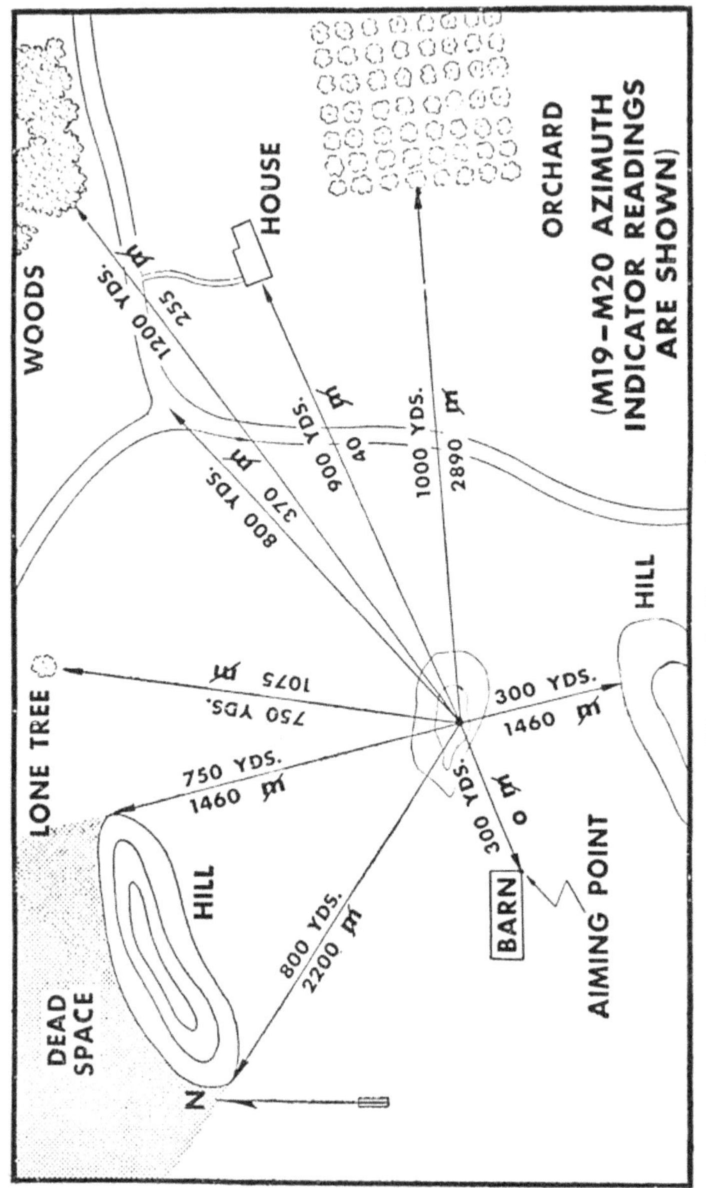

Figure 6. Range card.

c. If the members of a gun crew become casualties, the range card permits the replacement crew to deliver effective fire on likely targets from that position.

17. PROCEDURE. Range cards are prepared in duplicate for each gun immediately upon occupation of a position. The crew commander, assisted by the other members of the crew, determines ranges to prominent terrain features and plots them on his range card. Ranges may be determined by any or all of the methods outlined in paragraph 7. When a target appears in the area, a quick and accurate determination of the range may then be made by referring to the nearest point on the range card.

18. CONSTRUCTION. A range card is constructed as follows:

a. Either drive into position or mark the position with a stake.

b. Select a terrain feature as an aiming point, or set out aiming posts or improvised stakes. When in position, lay on this aiming point and zero the azimuth indicator. If an aiming circle is being used, zero it on the aiming point.

c. At the approximate center of a blank piece of paper, place a dot for the gun position. Draw a line through this dot to another dot representing the location of the aiming point or stake.

d. Measure the angles to terrain features or probable targets. With the origin at the gun position, construct the angles thus determined from the gun position—aiming point line.

e. Determine the range to each object. Plot and sketch the objects on the proper lines and label them.

f. If the gun can be fired in all directions, plot points for all-around fire.

g. Write range and deflection on each line.

h. Change range estimates as more accurate data are obtained. Correct the ranges when the gun has been fired at any targets in the area.

i. If the position is to be used for night firing, carefully determine the range and angle of site, and write quadrant elevations on the line to each target. (See sec. III, ch. 14.)

j. Lay each gun in the approximate center of its sector. Use terrain features to define the limits of the sector to each crew member. Identify the terrain features and targets, to which deflections have been recorded, to all members of the crew.

k. To provide coordination and allow massing of fires, assign each terrain feature or probable target a common designation within the platoon or company.

19. SPECIAL FORM. Some units have found it desirable to prepare range cards of the design shown in figure 7. After this form is printed, it is glued onto a piece of masonite or similar material and covered with acetate or clear lacquer. The range card is drawn on the acetate or lacquer. After moving to a new position, the old drawing is erased and the range card for the new position is prepared on the same card.

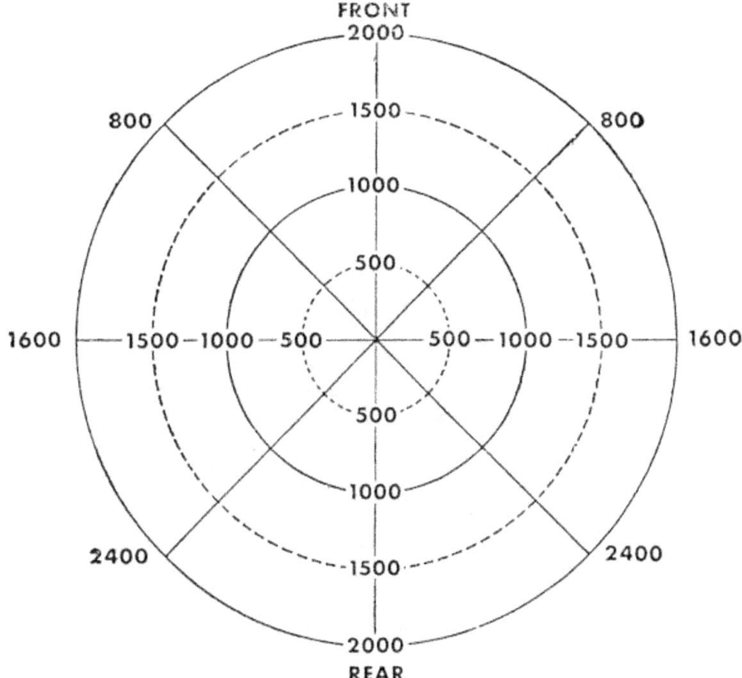

Figure 7. Special form range card.

CHAPTER 3

SPEED DETERMINATION

20. GENERAL. *a.* In moving target fire, the effectiveness of the first round depends upon speed as well as range determination. The ability of the crew commander to estimate speed determines the correctness of the initial lead.

b. The rules for range estimation instruction (par. 8) apply to training in speed estimation. The types of vehicles used in training and the terrain over which they move should be varied. Practice in estimation of speeds should be conducted when vehicles are moving on roads and when they are moving cross country.

c. The following exercises should be used repeatedly:

(1) Tanks and other vehicles are driven across the line of observation at various ranges. The speed at which the vehicles are traveling and the lead necessary at each speed are announced.

(2) Vehicles appear at various ranges, moving at prearranged speeds. The class is required to estimate the speed and range of each and to announce the range and the necessary number of leads.

21. ESTIMATION OF SPEED. *a.* The speed at which the target is moving perpendicular to the direction of fire determines the amount of lead. To determine this factor, the crew commander establishes a line of sight parallel to the tube. He then estimates the target's speed of movement away from his line of sight. He is not concerned with the speedometer speed of the vehicle, but only with its speed away from his line of sight. This speed is referred to as "apparent speed." In figure 8, vehicle *A* may have a speedometer speed of 30 miles per hour, but it has no apparent speed and, therefore, no lead is required. Vehicle *B*'s speed, however, is carrying it away from the line of sight; therefore, it must be given a lead commensurate with its apparent speed.

b. Speeds are estimated as follows:

Very slow (1–5 mph)
Slow (6–10 mph)
Medium (11–20 mph)
Fast (21–30 mph)
Very fast (over 31 mph)

c. The horizontal lines and the spaces between them on the sight reticle establish angular leads in units of 5 mils. Since the lead is angular, the range to the target need not be considered in determining the lead. Taking one lead at 500 yards points the gun 2½ yards in advance of the center of the target. Taking one lead at 1,000 yards points the gun 5 yards in advance of the center of the target. A target moving 8 mph would move these same amounts at the ranges specified during the time of flight of a 76-mm APC, M62 projectile. Therefore the range does not affect the angular lead.

d. For a gun with a muzzle velocity of approximately 2,600 feet per second, the following num-

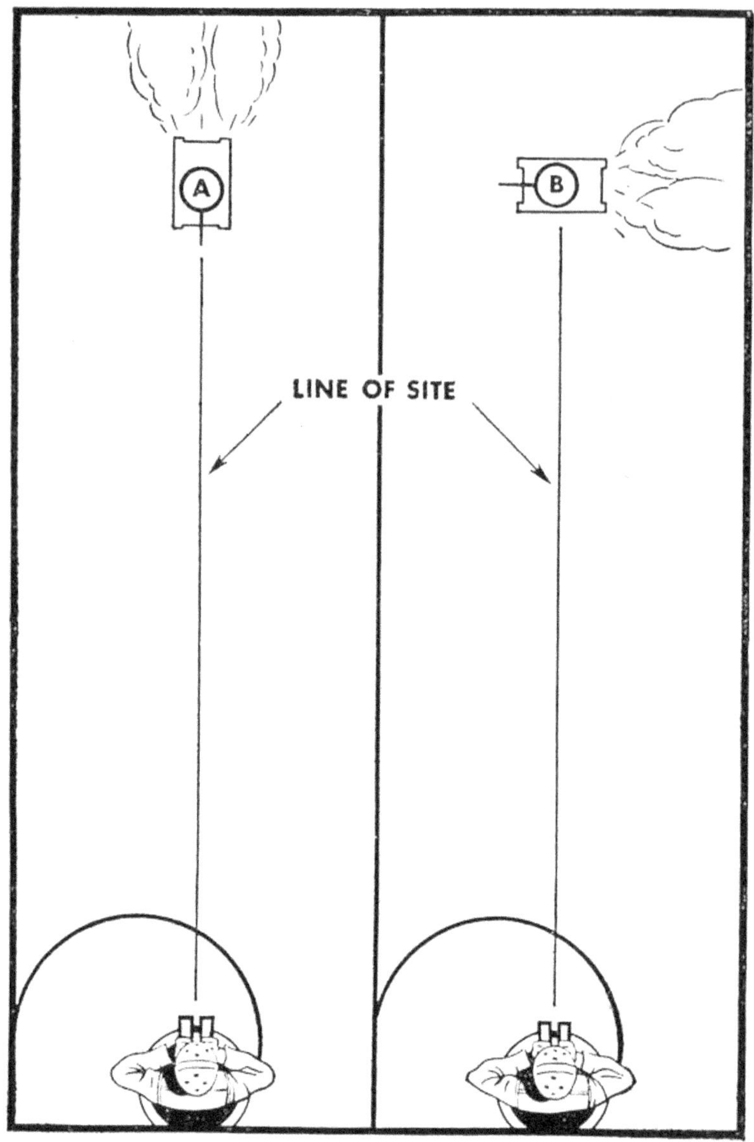

Figure 8. Determining apparent speed.

ber of leads is required for the indicated rates of apparent speed:

Very slow	One half lead.
Slow	One lead.
Medium	Two leads.
Fast	Three leads.
Very fast	Four leads.

e. For a gun with a muzzle velocity of approximately 2,000 feet per second, the following number of leads is required for the indicated rates of apparent speed:

Very slow	One half lead.
Slow	One lead.
Medium	Two leads.
Fast	Four leads.
Very fast	Five leads.

f. For a gun with a muzzle velocity of approximately 1,500 feet per second, the following number of leads is required for the indicated rates of apparent speed:

Very slow	One half lead.
Slow	One lead.
Medium	Four leads.
Fast	Six leads.
Very fast	Eight leads.

CHAPTER 4

ANGLE OF SITE

22. GENERAL. A projectile fired at the firing table elevation for a range of 1,000 yards would hit a target placed at 1,000 yards if the target were

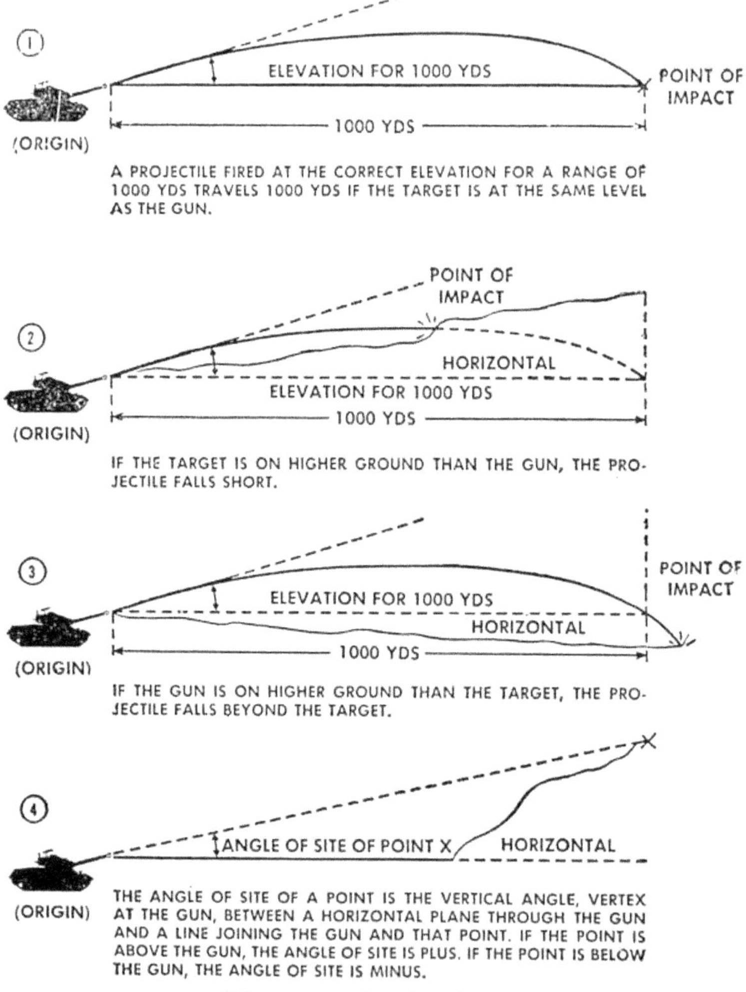

Figure 9. Angle of site.

at the same level as the gun (fig. 9) and if all conditions affecting the trajectory were normal. If the target is on higher ground than the gun, the projectile will fall short. If the target is lower than the gun, the projectile will fall beyond the target. This difference in altitude must be compensated for when the sight reticle is not used to lay for range. It is reduced to an angular measurement known as the "angle of site" and is applied as a correction to the elevation for the range.

23. DEFINITION. The angle of site of a point is the vertical angle at the gun between the horizontal and a line joining the point and the gun. If the point is above the gun, the angle of site is *plus*. If the point is below the gun, the angle of site is *minus*.

24. EFFECT ON TRAJECTORY. **a.** When the angle of site is applied to a gun, the entire trajectory is rotated about the origin as though the trajectory were a rigid curve. (See fig. 10.)

b. Laying the reticle directly on the target applies the angle of site when the proper range line is laid on the target.

25. DETERMINATION. **a.** When firing against point targets, the angle of site must be determined with extreme care. With the 76-mm gun, an error of 5 mils in angle of site at 2,000 yards is equivalent to an error of 500 yards in range.

b. When the gunner can identify the target and

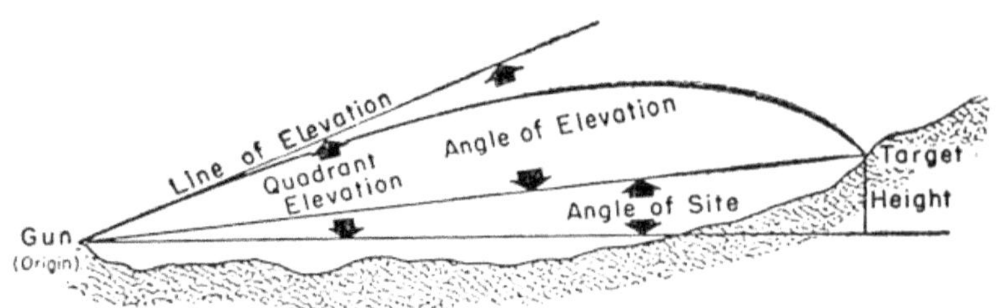

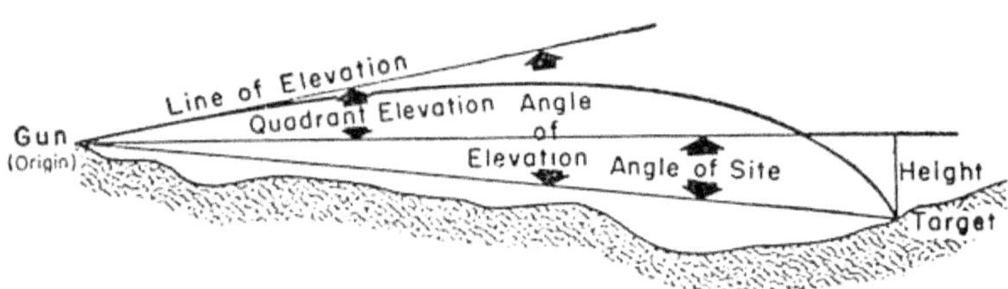

Figure 10. Rotation of trajectory by applying angle of site.

it is necessary to determine the angle of site, the following method can be used:

(1) Lay the boresight cross of the coaxial telescope on the target.

(2) The elevation of the tube is determined with the gunner's quadrant or elevation quadrant (See ch. 9.) This is the most accurate method of determining the angle of site.

c. When the observer's position is at an altitude which differs considerably from that of the gun, it may be necessary to compute the angle of site accurately. The procedure is prescribed in FM 6–40.

d. When the observer's position is close to the

gun, one of the following three methods may be used:

(1) (*a*) Measure the angle of site with an aiming circle by laying the *vertical* cross hair slightly to the right of the target.

(*b*) Level the angle of site bubble.

(*c*) Read by interpolation on the vertical mil scale the number of mils from the horizontal cross hair to the target. Above the horizontal cross hair, the value is *plus;* below, it is *minus.*

(2) When shifting from one target to another target which is at approximately the same range but at a different altitude—

(*a*) Measure the vertical angle between the two in mils with field glasses or an aiming circle.

(*b*) Apply the angle thus determined to the site used for the original target.

(3) When speed is important—

(*a*) Pick a distant point on approximately the same level as the gun.

(*b*) Read the vertical difference in mils on the field glass reticle between that point and the target. (See fig. 11.)

(This method is comparatively inaccurate.)

26. APPLYING ANGLE OF SITE. a. When laying with the reticle, no command for site is necessary. Laying the proper range line on the target automatically includes angle of site.

b. In vehicles equipped with the graduated handwheel, apply the angle of site by means of the graduated handwheel after laying for range with the elevation quadrant. (See ch. 9.) Announce

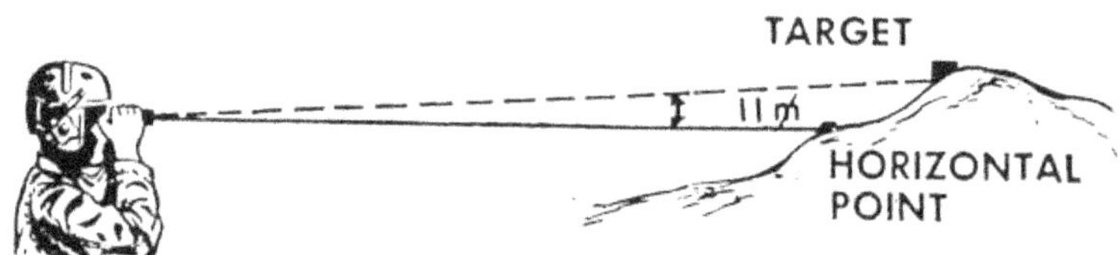

Figure 11. Determining angle of site with field glasses.

the range followed by the angle of site in mils. For example, if the site is plus 5 mils: TWO THOUSAND, UP FIVE.

c. In vehicles not equipped with graduated handwheels, add a plus angle of site to or subtract a minus angle of site from the firing table elevation for the announced range. The total is announced as QUADRANT______, and is set off on the elevation quadrant or gunner's quadrant.

CHAPTER 5

CREW DRILL AND SIMULATED FIRE

27. GENERAL. a. Accuracy must be stressed. It can be developed by constant checking and by insistence on attention to detail. Inaccuracies permitted during training will multiply in battle.

b. Speed should be developed but never by the sacrifice of accuracy, to which it is always subordinate. Do not seek it before the state of training warrants. Speed is obtained by constant practice and by the elimination of lost motion.

c. Each training period should be planned in advance. Crew drill and simulated firing exercises become extremely tiresome and defeat their purpose if periods are too long. To maintain proficiency, drills should be held every day although the periods are short.

d. In all firing and simulated firing exercises, the crew commander designates the target and gives the proper fire orders. Each member of the crew performs all his duties, as prescribed for the service of the piece. The interphone system is used except when a secondary means of intracrew communication is being taught.

e. To obtain accurate fire, sights must be in adjustment. Sight adjustment will be checked before each period of simulated firing. This will enable crew members to improve their boresighting ability.

f. Settings and layings should be checked frequently. Checking is most valuable when unexpected.

28. SEQUENCE OF INSTRUCTION. **a.** The crew members should be shown their posts, the correct method of mounting and dismounting from the vehicles, the individual duties during firing, and the proper stowage as prescribed in appropriate field and technical manuals. This training may be interspersed with training in sight pictures, laying, and conduct of fire.

b. Each member of the crew must be trained to perform the duties of all of the other members. Gunner training should include ample practice with the gunner trainer (par. 114), in manipulation (par. 115), and in laying with the quadrant and graduated handwheel (ch. 9). Training of the crew commander should include exercises on the terrain board, the terrain fan, the blackboard, and in the firing of problems on the fire adjustment trainer. (See pars. 119–121.)

c. After elementary crew drill, instruction should be given in individual duties. This will be followed by instruction in the duties of teams within the crew (commander and driver, commander and gunner, gunner and loader), and, finally, of the entire crew. The importance of the driver in gunnery must not be overlooked. The necessity for teamwork will be constantly emphasized.

d. Simulated firing should be combined with driving instruction and simple maneuvers. Targets will be tanks, other vehicles, and terrain objects at ranges known to the instructor but not to the crews. The types of targets should be varied to provide training in methods of adjusting both shot and HE. Shooting should begin with known-distance targets to obtain accuracy, and turn quickly to targets at unknown ranges. Known-distance firing is not realistic.

e. Training includes simulated firing at stationary and moving targets. Whether or not the vehicles are equipped with gyro-stabilizers, the crews should also practice firing machine guns while moving. (See sec. V, ch. 14.)

29. INDIVIDUAL DUTIES. a. Platoon leader. The platoon leader is responsible for the complete training of the entire crew. He deals with the crew through its commander.

b. Crew commander. During crew drill, the commander watches every movement of his men. He does not interfere except when his assistance is required or when he observes an error. He should never permit observed errors to go uncorrected.

c. Gunner. Accuracy and speed are the principal qualities to develop in a gunner. Accuracy is developed first and then is combined with speed. The following common faults must be detected and eliminated:

(1) Failure to lay exactly on the target.

(2) Failure to verify the laying for direction and elevation for each round after the breech is closed.

(3) In indirect fire, failure to lay always on the same part of the aiming point or aiming post.

(4) Failure to take up the lost motion in the traversing and elevating gears when firing at a stationary target. The gunner takes up lost motion in the elevating and traversing gears by making the final movement of the piece in the direction in which it is more difficult to elevate or traverse. When there is no direction of greater resistance, the last movement in elevation and the last movement of traverse are always in the same direction as the initial movements. It may often be necessary to traverse or elevate beyond the target and then come back on it. In gun carriages equipped with the gyro-stabilizer, turret guns are balanced within a few pounds for stabilizer operation. When the stabilizer is not operating, a track block hung on the recoil guard makes the gun sufficiently breech heavy to insure that all lost motion in the elevating gears can be taken up.

(5) Failure to bring pointers into *exact* alignment with index marks when using the graduated handwheel or azimuth indicator.

(6) Failure to level bubbles exactly.

(7) Failure to continue tracking at the instant of firing.

d. Loader (assistant gunner). The loader wipes off the ammunition before loading. He holds the next round to be loaded away from the path of recoil. He must be taught to insert the round smoothly into the breech recess and to push it home with sufficient impetus to seat it in the chamber. Timid loading, caused by fear of getting fingers caught in the breechblock, results in failure of the breech to close and may cause a jammed round. Every effort should be made to force such a round home before attempting to extract it. Extraction will usually result in a separated round.

30. PEP DRILL. Crew drill and simulated firing exercises become dull and lax unless the crews are alerted by unexpected periods of pep drill, as explained in pertinent field manuals.

31. TRAINING. For additional information on crew training, see chapter 13.

PART TWO

TECHNIQUE OF GUNNERY

CHAPTER 6

GENERAL

32. DEFINITION. The technique of gunnery includes the principles involved and the procedure used in the delivery of effective fire upon a target.

33. RESPONSIBILITIES. To deliver effective fire accurately and promptly on a target, the following individuals must assume the responsibilities listed.

a. Platoon leader. The fire power of the platoon is most effective when it is coordinated by the platoon leader. He should designate specific targets or sectors of responsibility to the guns of his platoon and may indicate when each gun or section is to open or cease fire. In this manner, he insures coverage of all important targets and prevents waste of ammunition on unimportant targets.

b. Crew commander. The crew commander is responsible for the actions of his crew. He gives initial fire commands, determines corrections, and gives subsequent commands.

c. Gunner. The gunner lays the gun for deflection and elevation or range. He executes the fire orders given by the crew commander. When ready to fire, the gunner alerts the crew commander by announcing ON THE WAY. This warns the crew commander to raise his glasses and to be ready to observe.

34. TRAINING. Direct fire gunnery training includes the following subjects which will be treated individually in succeeding chapters:

a. Fire orders.

b. Marksmanship, knowledge of the sight reticle, sight adjustment, and initial and subsequent laying.

c. Conduct of fire with shot, including an understanding of trajectory and vertical dispersion.

d. Conduct of fire with HE, including an understanding of range dispersion.

e. Additional methods of direct fire.

CHAPTER 7

DIRECT FIRE SIGHTS

Section I. GENERAL

35. TRAINING. A thorough knowledge of the sight, its use, and its adjustment is the basis of good gunnery. Every gunner must understand completely the capabilities and limitations of his sights. High velocity guns are designed to effect penetration of vertical targets with a minimum of adjustment. To use these weapons to best advantage, gunners must be trained to pin-point accuracy.

Section II. SIGHT RETICLE

36. DESIGN. The sight reticle is designed to permit the gunner to lay accurately on direct fire targets for both range and deflection. The reticle can be best explained to the new gunner by showing him how the design of the reticle includes both of these elements.

37. CONSTRUCTION. A model reticle may be constructed in the following manner:

a. The field of view of the sight is represented by a large circle.

b. The boresight cross is placed in the reticle to establish a line of sight. The intersection of these lines is zero range and zero deflection. The bottom of the vertical line is the 200-yard range mark. (See fig. 12.)

c. The 400–600-yard tick is added, then the 800–1,000-yard tick, etc. (The number of ticks will vary with the particular sight under consideration.) Each range tick represents 200 yards

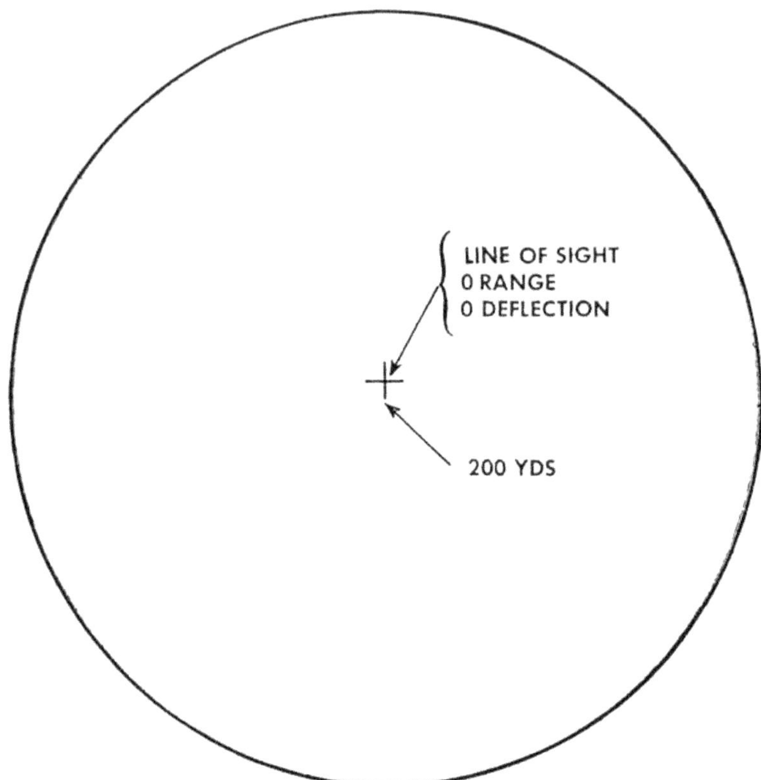

Figure 12. Line of sight.

and each interval between ticks is also 200 yards. The length of the range ticks increases as the range increases. Thus, the 4,000–4,200-yard tick will be longer than those for shorter ranges. (See fig. 13.)

d. The first 5-mil deflection or left lead line is placed in position at the 400-yard graduation.

(See fig. 14.) This line is 5 mils long. The space separating it from the vertical line is also 5 mils.

e. The second and third deflection lead lines are added on the left side of the vertical axis. (See fig. 15.) This is continued until the left side of the reticle is complete.

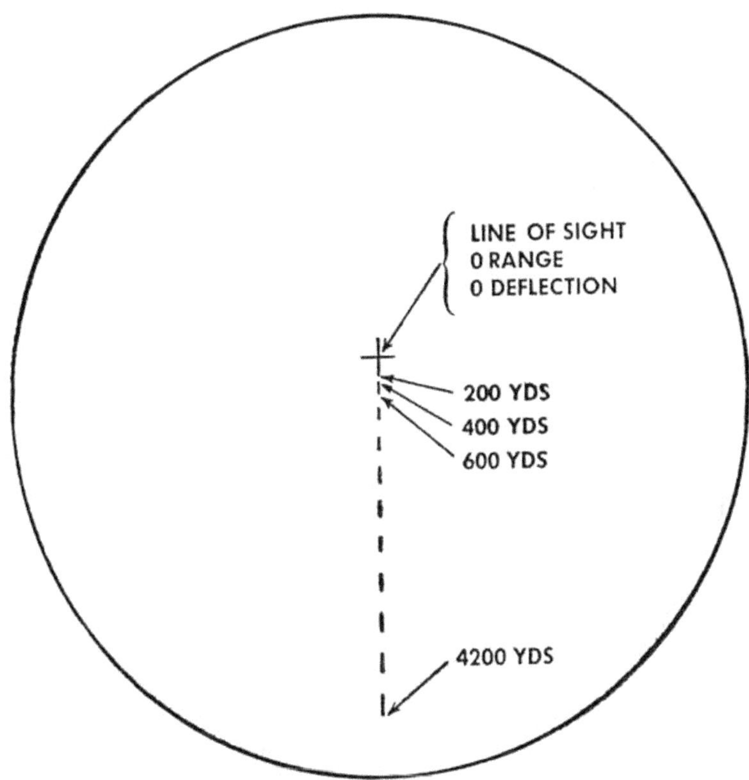

Figure 13. Range ticks.

f. After all deflection lead lines and numerals designating the various ranges have been placed, the left side of the reticle will appear as shown in figure 16.

g. Similarly, right deflection lead lines will be

added, completing the reticle. Each sight reticle is graduated for a particular type of ammunition. The type for which it is graduated is marked near the top of the field of vision. (See fig. 17.)

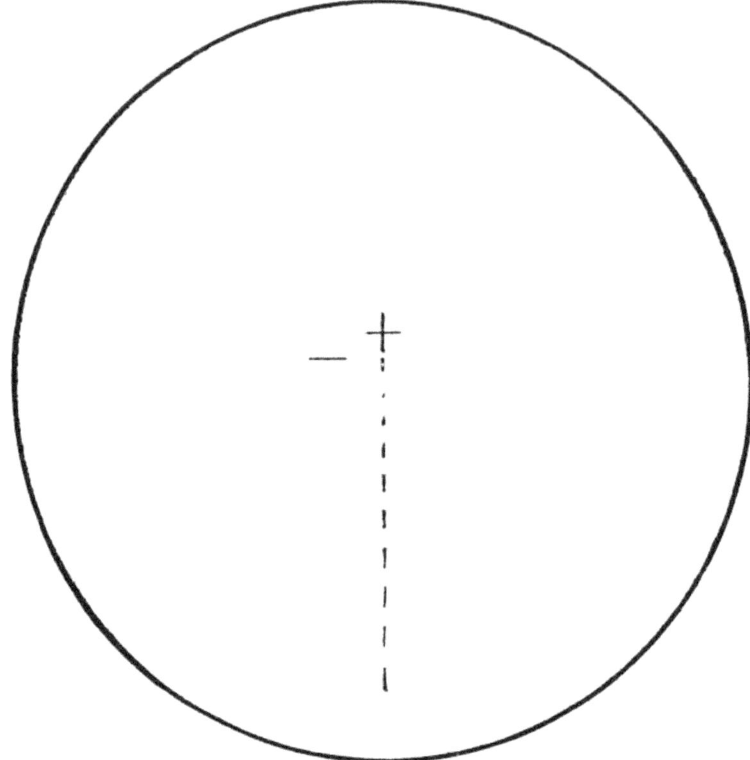

Figure 14. First left deflection-lead line.

h. The same steps may be followed in building up any type of reticle.

38. DRAWING RETICLE. After completing the steps in paragraph 37, each gunner should draw a picture of the sight reticle under the supervision of his crew commander, section or platoon leader, and should label the ranges.

39. INITIAL LAYING WITH SIGHT. **a.** Direct fire sights are graduated so that when the correct range graduation is placed on the center of the target, the gun tube will be elevated to the required quadrant elevation.

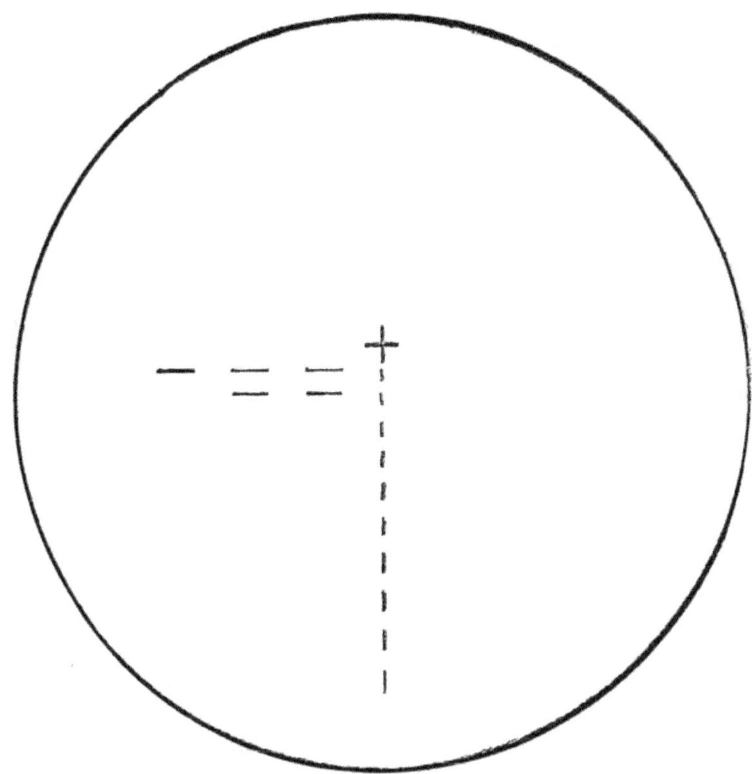

Figure 15. Second and third deflection-lead lines.

b. This is true, however, only when using the ammunition for which the sight is calibrated. The gunner must compensate when using any ammunition other than the one specified on the sight reticle. The aiming data chart (figs. 18 and 19) is used to make this correction. The aiming data charts illustrated are for the 76-mm gun on Motor Carriage, M18, and the 90-mm gun on the Heavy Tank, M26. Similar charts can be prepared for any gun and mount by a comparison of firing

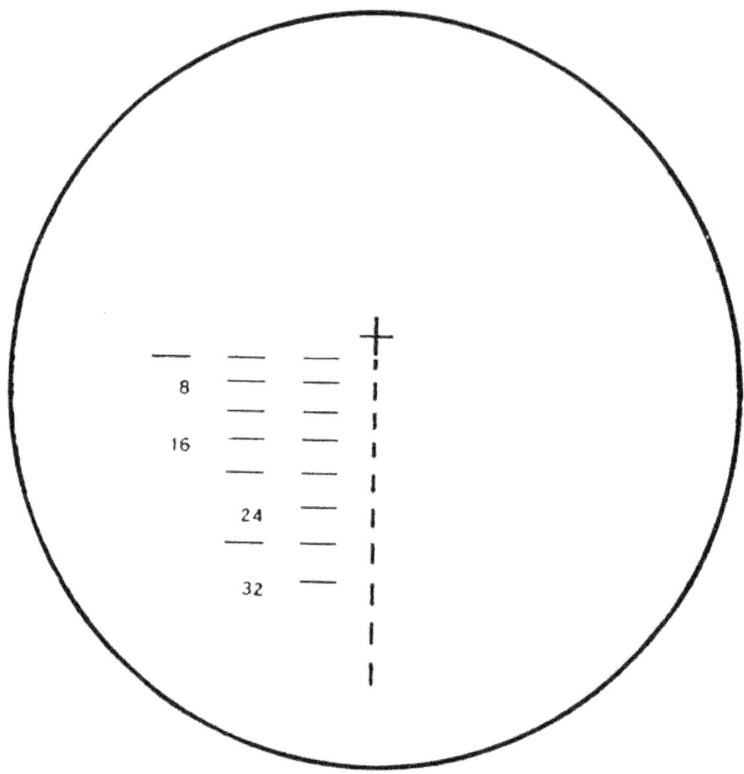

Figure 16. Left side of sight reticle.

table elevations for the different types of ammunition.

c. To lay initially, the intersection of the proper range line and the center vertical line of the reticle is placed on *the center of mass of the visible portion of the target.* At very short ranges, the aiming point may be shifted to the most vulnerable spot which is visible, such as the embrasure of a pillbox or a lightly armored point on a tank.

d. The horizontal lines of the sight reticle are used—

(1) To establish leads when firing at moving targets.

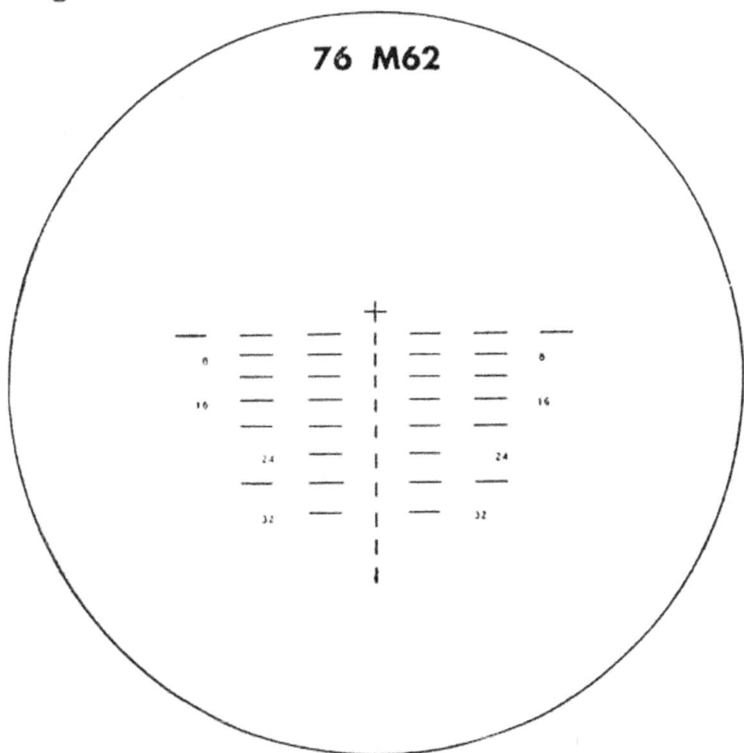

Figure 17. Completed sight reticle.

(2) To lay for deflection when firing at stationary targets which are identifiable by the gunner.

40. LAYING WITH LEADS. a. To hit a target which has any movement perpendicular to the direction of fire, it is necessary to aim ahead of the target. This process is referred to as "leading the target."

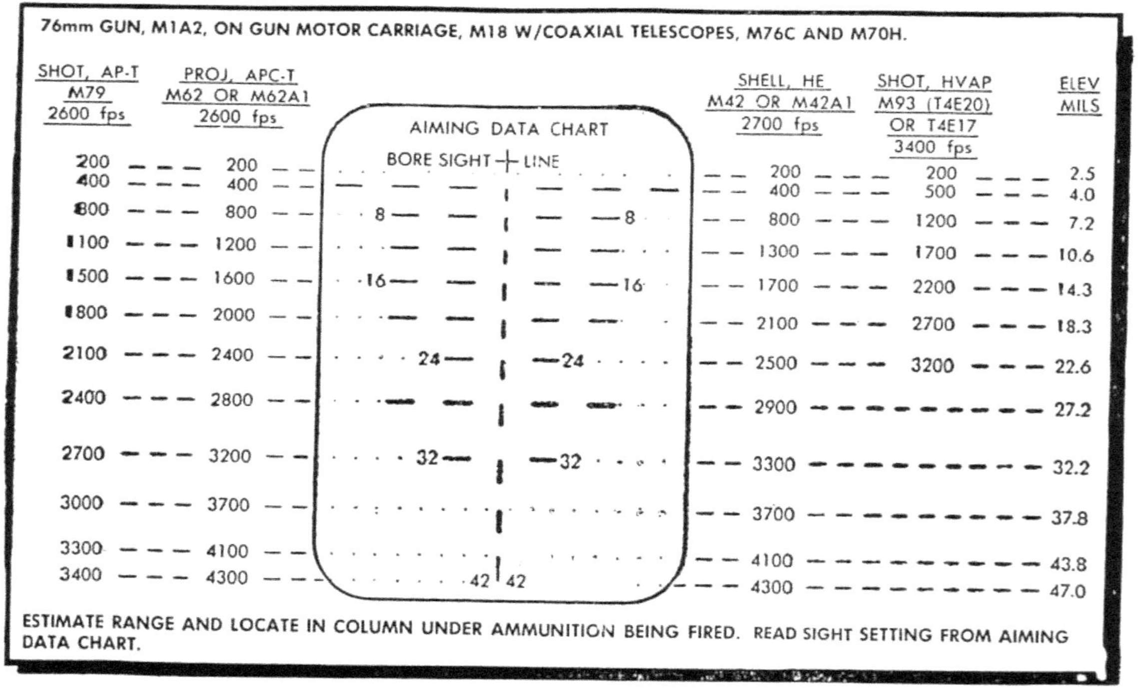

76mm GUN, M1A2, ON GUN MOTOR CARRIAGE, M18 W/COAXIAL TELESCOPES, M76C AND M70H.

SHOT, AP-T M79 2600 fps	PROJ, APC-T M62 OR M62A1 2600 fps	AIMING DATA CHART (BORE SIGHT LINE)	SHELL, HE M42 OR M42A1 2700 fps	SHOT, HVAP M93 (T4E20) OR T4E17 3400 fps	ELEV MILS
200	200		200	200	2.5
400	400		400	500	4.0
800	800	8	800	1200	7.2
1100	1200		1300	1700	10.6
1500	1600	16	1700	2200	14.3
1800	2000		2100	2700	18.3
2100	2400	24	2500	3200	22.6
2400	2800		2900		27.2
2700	3200	32	3300		32.2
3000	3700		3700		37.8
3300	4100		4100		43.8
3400	4300	42	4300		47.0

ESTIMATE RANGE AND LOCATE IN COLUMN UNDER AMMUNITION BEING FIRED. READ SIGHT SETTING FROM AIMING DATA CHART.

Figure 18. Aiming data chart.

90MM GUN (W/MUZZLE BRAKE) M3 ON HEAVY TANK, T26E3, W/COAXIAL TELESCOPE, M71C AND PERISCOPE TELESCOPES, M77F AND M80D.

SHOT, AP-T T33 2800 FPS	SHOT, HV, AP-T T30E15 3350 FPS	PROJ, APC-T M82 2800 FPS	AIMING DATA CHART (BORE SIGHT LINE)	SHELL, HE M71 2700 FPS	PROJ, APC-T M82 2670 FPS	ELEV MILS
300	400	200		100	100	1.6
500	700	400		400	300	3.1
900	1200	900	8	800	700	6.0
1300	1800	1300		1200	1100	9.2
1700	2300	1700	16	1600	1500	12.4
2200	2700	2200		2100	1900	15.9
2600	3200	2600	24	2500	2300	19.5
3000		3000		2900	2700	23.4
3400		3500	32	3300	3100	27.5
3800		3900	36	3700	3500	31.8
4200		4300	40	4100	3900	36.5
4600		4800		4500	4400	38.9
4800		5000		4700	4600	41.4

ESTIMATE RANGE AND LOCATE IN COLUMN UNDER AMMUNITION BEING FIRED. READ SIGHT SETTING FROM AIMING DATA CHART

Figure 19. Aiming data chart.

b. Angular leads are used for moving target firing. To give the gunner a scale for applying leads, the sight reticle is provided with a horizontal scale graduated in 5-mil units.

c. When laying with leads (fig. 20), the vertical line must be kept ahead of the center of mass of the target.

d. The proper range line is laid on the point of

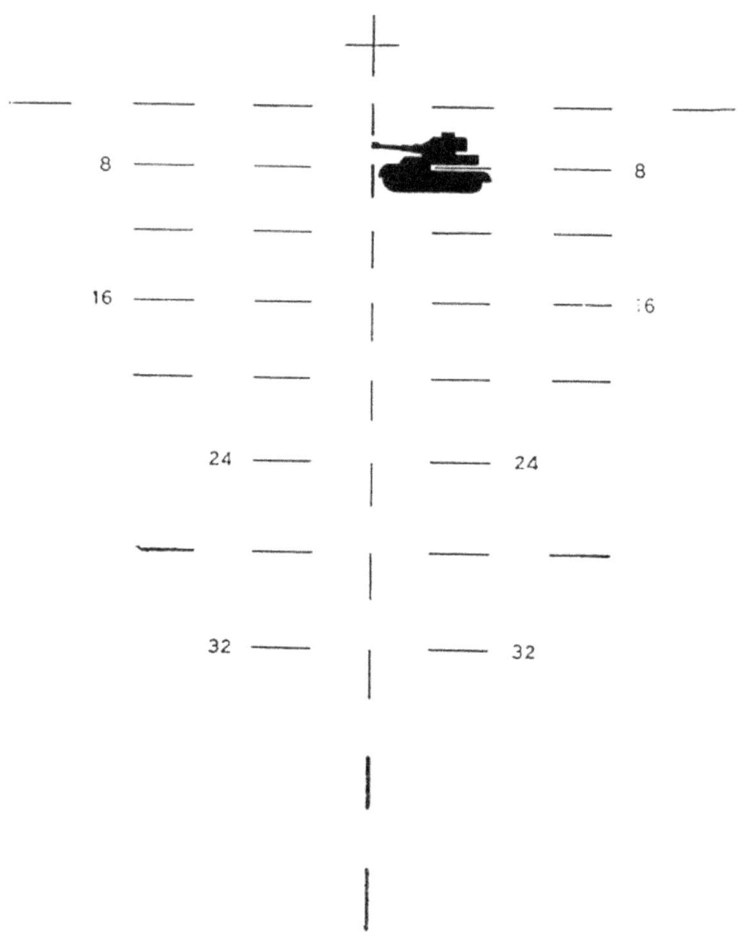

Figure 20. Laying for lead (One Lead, 800 yards).

aim. The piece is traversed smoothly through the target from rear to front and is swung ahead to the announced lead. Tracking is done smoothly. All changes in lead are made gradually. *Tracking is continuous while firing* at a moving target. Gunners will not lay the gun ahead of the target and wait for it. Gunners must not flinch while firing but must continue to track the target, adjusting lead and range as directed by the crew commander. Adjustments are made from the sight picture at which the last round was fired.

41. LAYING FOR DEFLECTION. When firing at stationary target at ranges in excess of 1,500 yards, the factors of drift, wind, and cant must be considered.

a. Drift is the deviation of the shell from the vertical plane through the axis of the tube caused principally by the rifling in the bore. This drift is to the right for all weapons discussed in this manual. The drift effect may be found in appropriate firing tables. Some sights are designed to compensate for drift. (See fig. 21.)

b. The effect of a crosswind is to deflect the projectile from the standard trajectory. The effect of a lateral wind of 1 mile per hour can be found in the appropriate firing tables.

c. Cant is the inclination of the trunnions of the gun from the horizontal.

(1) Cant is caused by failure to place the gun carriage on level ground. Cant of the gun should not be confused with "apparent cant" which is caused by a defective or improperly adjusted sight.

(2) A moderate amount of cant does not cause any appreciable range error, but does introduce a deflection error. To correct for cant in laying initially for direction on a stationary target, drop an imaginary vertical line from the boresight

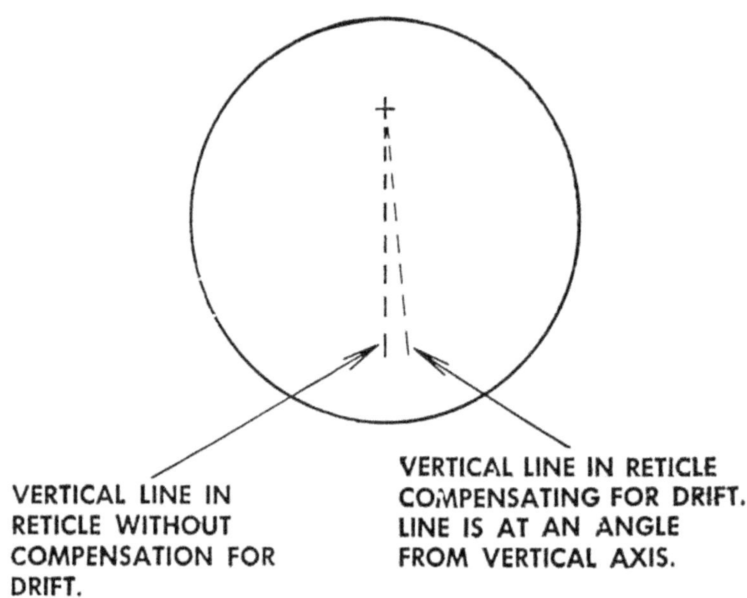

Figure 21. Sight compensated for drift.

cross on the reticle and lay this line on the target. (See fig. 22.) After the first round, fire is adjusted by observation.

(3) Cant is eliminated by leveling the trunnions. This can be done by either selecting or making a level firing position for the carriage. When fire is to be delivered in one direction only, it will suffice to level the trunnions perpendicular to that direction.

(4) "Apparent cant" is prevented by careful

handling of sights and by removal of all play from sight mounts. To detect "apparent cant" caused by a faulty reticle or sight bracket, make sure that the trunnions are level, lay on a point

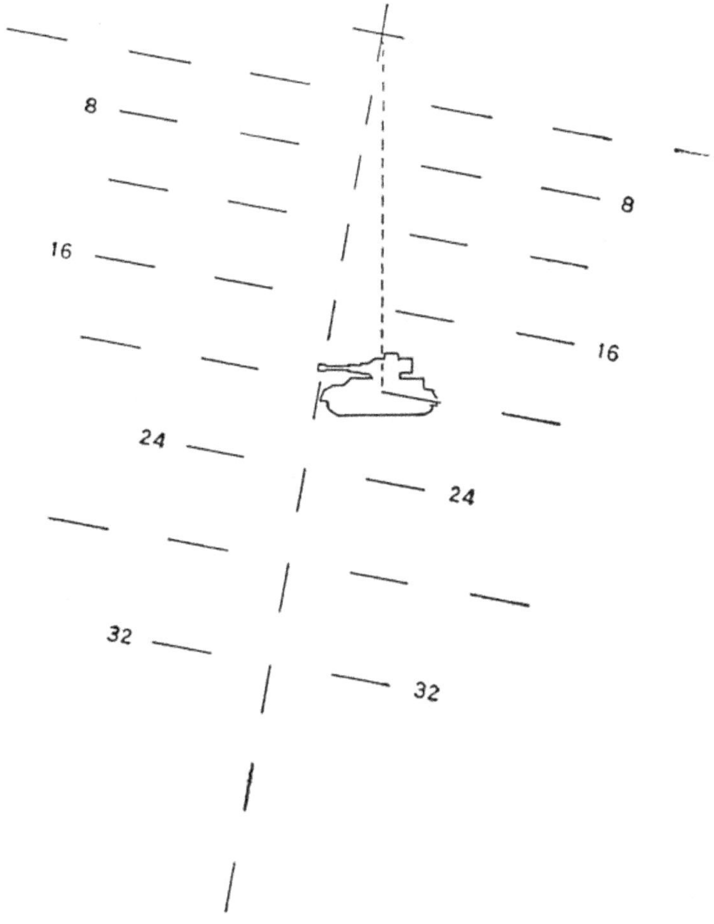

Figure 22. Canted sight reticle.

target, and then slowly elevate the gun. If the vertical line of the sight reticle remains on the target, the sight has no cant. If the target moves to the right or left of this vertical line, the sight reticle or bracket may be defective. This fact should be reported to Ordnance. In making this test, it must be remembered that some reticles are designed to compensate for drift. (See fig. 21.) In checking these reticles, "apparent cant" is present if the angular displacement between the vertical line and the target differs in direction or amount from the firing table drift for the sight range at the point of comparison.

42. SUBSEQUENT LAYING WITH SIGHT. **a.** The only individual who knows exactly how the piece was laid at the instant it was fired is the gunner. In direct fire, whether the target is stationary or moving, the gunner continues to look through the sight while the piece is being fired. Thus, he has an exact picture of the relationship between the target and the sight reticle. Just as the rifleman is taught to call his shot, so the gunner must know the exact laying at the time of firing.

b. Adjustments are announced as corrections or changes to be applied by the gunner to the *last sight setting.*

Section III. SIGHT ADJUSTMENT

43. GENERAL. Two methods are prescribed for sight adjustment, either by field boresighting or by zeroing. All sights, periscopes, spare heads and coaxial telescopes are adjusted. On tanks, the vane sight is also adjusted. The vane sight is adjusted by loosening the cap screws and aligning it for direction only.

44. FIELD BORESIGHTING. In field boresighting, the axis of the bore and the line of sight established by the boresight cross are made to intersect at an aiming point preferably in excess of the greatest range of employment and never less than the average range of employment, or at approximately 1,500 yards, if neither of these is known. The procedure is as follows:

a. The gun carriage should be on ground as level as possible.

b. Those parts of the firing mechanism necessary to open the firing pin well are removed. Issue breech boresights, if available, may be used.

c. Thin black cross hairs are placed on the muzzle so that they intersect at the exact center of the bore. This is accomplished by alining them with the witness lines on the muzzle.

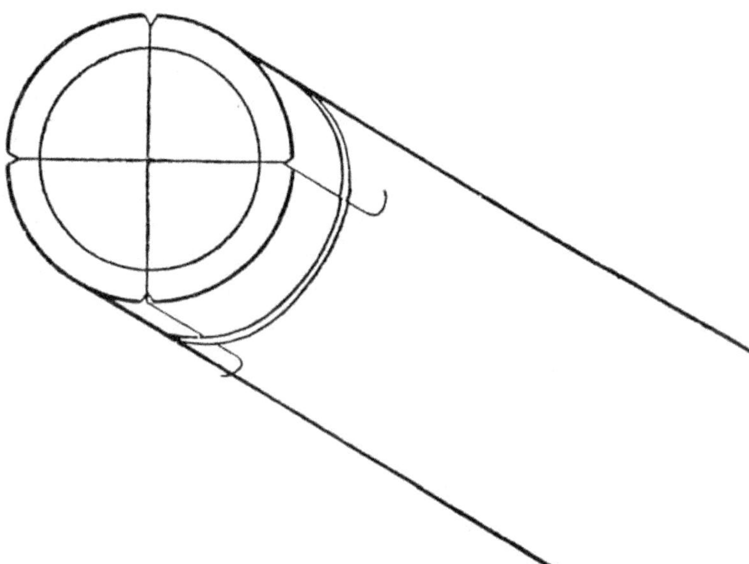

Figure 23. Thin black thread is stretched across the witness lines (reference grooves).

d. An object should be selected for sharpness and high contrast.

e. With the firing pin well as a rear sight and the cross hairs on the muzzle as a front sight, the axis of the bore is alined on a sharply defined spot on the aiming point. (See figs. 24 and 25.)

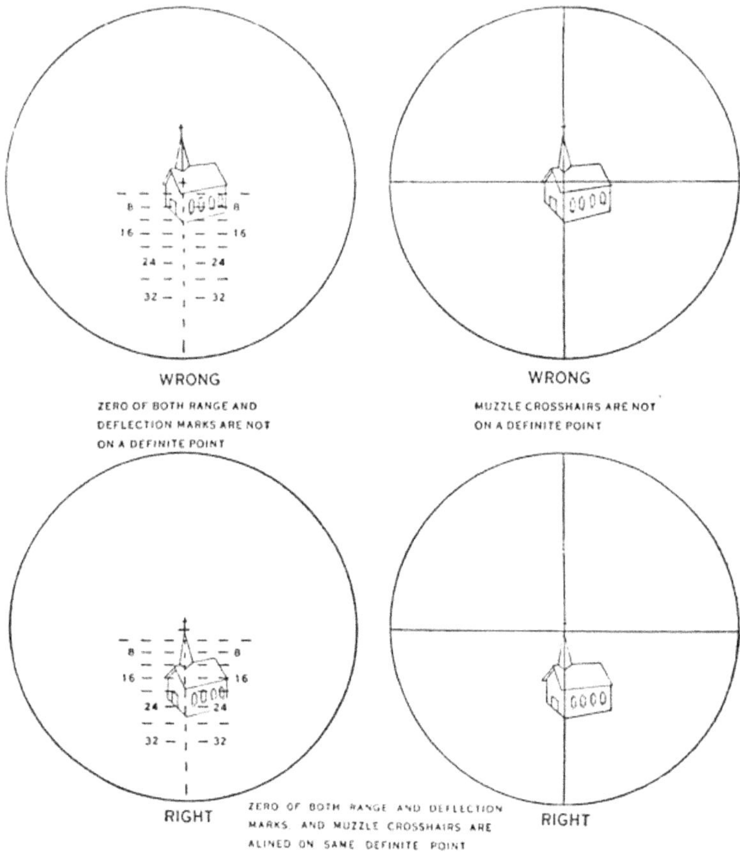

Figure 24. When boresighting, a definite point is selected.

With unfavorable light conditions, binoculars may be used advantageously to define the aiming point.

f. Without moving the tube, the sight is alined so that the intersection of the zero deflection line, that is the boresight cross, is laid on exactly the same point as the tube.

g. The sight should be clamped securely in position and both the tube and sight rechecked.

h. The gun is now boresighted.

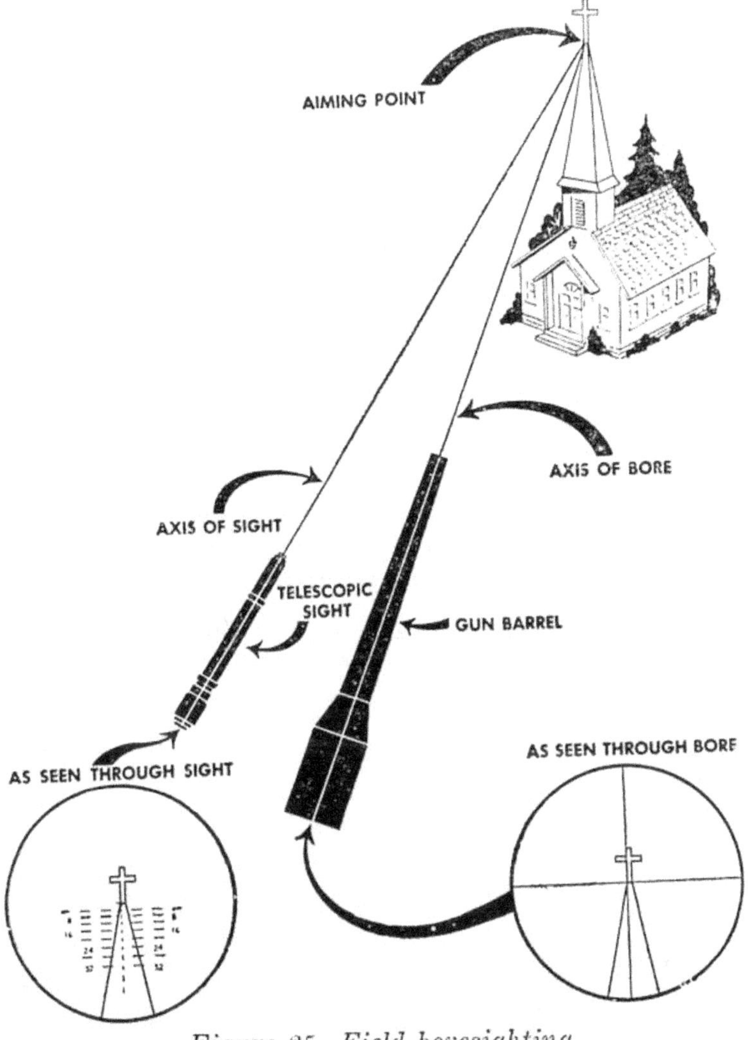

Figure 25. Field boresighting.

45. ZEROING. This is the most accurate method of direct fire sight adjustment and should be accomplished whenever practicable. The sights are adjusted so that the point of aim and the point of strike of the projectile coincide at the range of average employment, or if this is unknown, at approximately 1,500 yards. The procedure is as follows:

a. Field boresight. (See par. 46.)

b. Place an aiming cross on the center of a 12 x 12 foot panel.

c. Place the panel at a known range approximating that of average employment, or if this is unknown, at approximately 1,500 yards.

d. Without laying on the panel, fire a round to warm up the gun.

e. Using the aiming cross and the proper range graduation, fire three to five rounds into the panel, relaying with the same sight picture after each round.

f. Relay on the aiming cross.

g. Adjust the telescopic and periscopic sights so that the vertical line and the appropriate range line are on the center of the three-to-five-round shot group.

h. Tighten the locking nuts on the sights securely.

i. Check after locking to see that the sights remain on the center of the shot group.

j. Fire at least one, and preferably three, check rounds to verify the center of the shot group.

k. The gun is now zeroed.

l. Immediately after the gun has been zeroed, it

will be boresighted from five to ten times on the same sharply defined aiming point. The gunner will record the sight picture, that is, the relationship between the aiming point as seen through the bore of the gun and the boresight cross, after each time he boresights the gun.

m. The center of the sight picture group is determined and recorded for reference in future checking of the sight adjustment.

46. CHECKING SIGHT ADJUSTMENT. Coaxial sight mounts provide sufficient stability to make unnecessary frequent or daily sight readjustment but a method of checking the adjustment is necessary. The procedure is as follows:

a. If the adjustment has been made by field boresighting, the adjustment is checked by another field boresight.

b. If the gun has been zeroed, the gunner obtains a boresight picture group by the procedure described in paragraph 45 l. The gun is boresighted at approximately the same range as it was zeroed.

c. The center of the sight picture group is determined.

d. If the center of the sight group is in satisfactory agreement with the center of the sight picture group obtained after zeroing, the sight adjustment should not be changed.

e. This check should be accomplished with the coaxial sight. The periscopic sight should be checked for coincidence with it and brought in agreement if necessary.

f. If the check indicates that the coaxial sight has moved out of adjustment, the gun should be zeroed again.

g. If it is impossible to zero the gun again, the sight should be readjusted by adjusting the sight until the center of the sight picture group obtained by boresighting the gun from five to ten times is in agreement with the center of the group obtained immediately after the gun was zeroed.

Section IV. MISCELLANEOUS SIGHT ADJUSTMENT

47. ADJUSTING SPARE PERISCOPE HEADS. a. Spare periscope heads are adjusted separately. All heads in the vehicle, including spares and those on the periscopes, are checked with each periscopic sight. The knob adjustments required in using the head on each sight are noted and recorded, together with the periscope serial number on the label of each periscope head. The serial number is on the name plate on the side of the periscope.

b. After installing a new head on a periscope sight, the gunner resets the adjustment knob according to the readings marked on the head for that sight. He must be sure to use the readings corresponding with the serial number of the periscope on which the head is then mounted. Each head label will have several readings on it, since it will have been adjusted with several different periscopes.

48. PLAY IN TELESCOPE MOUNTS AND PERISCOPE HOLDERS. Theoretically, if a telescope or periscope is removed from its mount or holder and is then replaced with the same setting on the adjustment knobs, the sight is still in adjustment. Practically, this is not likely to be true because of play in mounts and holders. After replacing a telescope or periscope, the sight should be readjusted. For the same reason, after a periscope has been adjusted it should not be pulled down to replace the cover over the adjusting knobs. In a loose holder, large errors may result.

49. TELESCOPE EYE SHIELD. There are various types of coaxial telescope eye shields. Three of these are adjusted as follows:

a. The soft rubber, bellowlike cup is the most common type eye shield. This is clamped to the diopter ring of the telescope. Light outside of the telescope is excluded by the pressure of the head with the eye centered in the eye shield opening. The rectangular brow pad directly above the eye shield is adjustable in depth. The gunner should position his eye to the eye shield and then adjust the brow pad as desired.

b. The firm rubber shield fits against the forehead and around both eye sockets acting both as a head rest and eye shield. The telescope fits through the right eye section of the shield and can be used successfully only by right eye dominant gunners. The shield is adjustable in depth to properly position the eye of the gunner to the eye piece.

c. The sponge rubber combination head rest and eye shield is the newest type eye shield. This shield is shaped to the contour of the eye socket and is fitted by a bracket around the telescope body. The shield is adaptable to either the right or left eye.

50. ADJUSTING SIGHTS FOR 1,000-INCH FIRING.

Unless offset targets are used for 1,000-inch firing, the abnormally short range and the miniature target make it necessary to converge the line of sighting and the axis of the bore at the target. This is done by adjusting the sight so that the point of aim and point of strike of the bullet coincide at this range. The procedure for zeroing the gun is as follows:

(1) Using the vertical cross hair and any desired range line in the sight reticle, the gun is laid on an aiming point on the 1,000-inch target and three single shots are fired. The gun is relaid after each shot.

(2) The center of the shot group is then determined. If the center of the shot group is on the aiming point, the gun is zeroed.

(3) If the center of the shot group is not on the aiming point, the gun is laid as before. Then without moving the gun, the sight is adjusted so that the vertical cross hair and the range line used in (1) above are laid on the center of the shot group in the target. In other words, the point of aim coincides with the point of strike.

(4) As a check, the gun is relaid on the aiming point and one round fired. If the above procedure has been followed accurately, the round should hit the aiming point.

(5) If an error still exists, steps (1) through

(4) above are repeated until the point of aim and the point of strike coincide.

b. Any range line may be selected for use in zeroing the gun.

c. On some gun carriages, periscopic sights are used for 1,000-inch firing because the coaxial telescope mounts do not have sufficient adjustment to permit zeroing at this range.

51. ADJUSTING SIGHTS FOR LONG RANGE SUBCALIBER FIRING. This may be accomplished by either zeroing or boresighting.

a. Zeroing subcaliber equipment. The subcaliber mount may be zeroed at a convenient range near the center of the range band to be fired. For example, if firing is to be between 600 and 1,000 yards, the gun should be zeroed at about 800 yards as follows:

(1) A clearly defined point at a known range of about 800 yards is selected. If the range is not known, it should be determined by the most accurate means available.

(2) Fire is adjusted until the selected point has been hit.

(3) Without moving the subcaliber gun, the telescopic sight is moved until the 800-yard range point of the zero deflection line is on the strike.

(4) The sight is clamped in position.

(5) A check round is fired.

(6) If an error still exists, the preceding steps are repeated until the point of aim and the point of strike coincide.

b. Boresighting subcaliber machine gun. (1) The

back plate and the bolt are removed from the machine gun.

(2) The boresight cross of the telescopic sight and the center of the bore of the machine gun are laid on a distant aiming point.

(3) The sight and the machine gun are clamped in this position.

52. OTHER ADJUSTMENTS OF PERISCOPIC SIGHT. **a** To check the adjustment of the linkage arm, proceed as follows:

b. Boresight on a distant aiming point. *While taking the following steps do not touch the sight adjustment again.*

c. Place the gun carriage on a steep forward slope and check the boresight when the gun is near its maximum elevation.

d. Place the gun carriage on a steep reverse slope which will bring the gun nearly to its maximum depression. Check the boresight again.

e. If the sights were properly adjusted with the gun carriage level but do not check accurately when the gun is elevated or depressed, the fault lies in the sight linkage. The linkage adjustment is second echelon maintenance.

f. The drag link between the gun trunnion and periscope holder is adjusted when installed so that the periscope accurately follows the elevation motion of the gun. If the link is too long or too short, the periscope will gain elevation or lose elevation as the gun is elevated. To verify the link adjustment, use a gunner's quadrant to measure the elevations of the gun and the periscope holder, taking similar readings at minimum and maximum elevation. If the link is correctly adjusted, the gun motion and the periscope motion will be equal.

g. The telescope of the periscope is adjusted when installed so that the line of sight through the center of the zero range line on the telescope reticle is parallel to the axis of the bore of the gun. The installation of a new head assembly on the periscope makes it necessary that the telescope be properly adjusted in the periscope. Do not remove the telescope or associated adjusting mechanism from the periscope body.

53. USE OF COLORED FILTER. **a.** The three colored filters now furnished with the M70 series telescopic sights are used at the discretion of the gunner.

b. Tests have determined that filters are an aid to clearer observation when used in the following manner.

(1) *Amber filter.* For observation of distant targets which are obscured by haze, especially targets which are bluish in color.

(2) *Neutral filter.* For intense light conditions.

(3) *Red filter.* For observation of targets obscured by smoke. (Use is limited to targets which have marked contrast with their background.)

CHAPTER 8

DIRECT FIRE ORDERS

Section I. GENERAL

54. INTRODUCTION. Crew commanders must be taught to give fire orders clearly and forcibly. Much time is lost when a crew member has to ask for a repetition of part of the fire order.

a. A forcible, confident tone of voice which conveys assurance to the crew should be developed.

b. When the crews have gained proficiency in service of the piece, the tempo of commands during drill periods is advanced until it is slightly faster than that which can be performed by the slowest man. This tends to speed up the slower crews.

c. Fire orders may be given by voice, voice relay, interphone, radio, or visual signals.

55. FIRE ORDERS. There are two types of direct fire orders:

a. Initial fire orders.

b. Subsequent fire orders.

Section II. INITIAL FIRE ORDERS

56. SEQUENCE. The following sequence is prescribed for the initial fire order:

Alert.
Type of ammunition.
Fuze (if necessary).

Direction.
Target description.
Range.
Leads (if necessary).
Command to open fire.

57. THE ALERT. The alert is always the first element of the initial fire order. It consists of the command FIRE MISSION to alert the crew followed by the individual who is to fire, as GUNNER or BOG (bow gunner). If it is desired to alert both, the command is GUNNER-BOG. When the platoon or section leader is controlling the fire of his unit, the alert also contains the number of the gun that is to fire, as NUMBER THREE.

58. AMMUNITION. The ammunition is designated as follows:

Armor-piercing or armor-piercing capped .... SHOT
Hyper-velocity armor-piercing (HVAP) .... HYPER-SHOT
High explosive .... HE
High explosive, antitank .... HEAT
Smoke .... SMOKE WP, SMOKE HC
Canister .... CANISTER
Caliber .30 machine gun .... CALIBER THIRTY

59. FUZE. When firing HE, the crew commander must specify the type fuze he wants; as DELAY, QUICK, or CONCRETE.

60. DIRECTION. Any of the following methods are suitable for designating direction. The method selected should be that which best suits the situation.

a. If the crew commander is in or on the vehicle, he may give—

TRAVERSE RIGHT (LEFT)
S–T–E–A–D–Y ------------------ON

b. When the crew commander is either inside or outside the vehicle, or when supporting infantry call for fire, any one of the following methods or a combination of them may be used:

(1) *General direction and clock system.* RIGHT FRONT, TWO O'CLOCK. "Front" is considered to be the direction in which the tube is pointing when the order is given.

(2) *Reference point and mils.* REFERENCE POINT, RIGHT TWO HUNDRED, or, REFERENCE POINT, LAST TARGET, LEFT NINE ZERO. Whenever time permits upon occupation of a position, a reference point near the center of the sector of fire should be designated. With azimuth indicator equipped vehicles, the gunner should lay on the reference point as soon as it is announced and should set the azimuth indicator at zero. This method is especially useful when the section or platoon leader is controlling the fire of his unit.

(3) *Reference point and distance.* REFERENCE POINT, LEFT 100 YARDS. This method should be used only at very short ranges.

c. A simple, rapid, and accurate method of designating indistinct targets is the use of a machine gun or the cannon. The person designating the target commands: WATCH MY TRACER (BURST). He then fires a round or a series of rounds at the target, and completes the designation orally. In using this method of designating the target, the crew commander should also designate the general direction of fire from his gun WATCH MY TRACER (BURST), MY DIRECT FRONT. The use of tracer ammunition in this method is limited by the tracer burn-out point of the ammunition used.

61. DESCRIPTION A brief description is all that is required, but it must create in the gunner's mind a picture of his target. If several targets are in view, the particular target on which fire is to be placed must be designated, as LEADING TANK or RIGHT BUILDING. The following words are used to describe the usual targets:

Any tank ---------------------- TANK
Armored car ------------ ARMORED CAR
Any unarmored vehicle ---------- TRUCK
Personnel ------------------- DOUGHS
Machine gun ----------- MACHINE GUN
Any antitank gun or artillery
piece ------------------- ANTITANK
Observation post -------------------- OP
Command post --------------------- CP

62. RANGE. The methods of determining range are discussed in chapter 2. When firing deliberately at stationary targets, ranges are announced to the nearest 50 yards. For moving targets, ranges are announced to the nearest hundred yards. Announce numbers as:

50	--FIVE ZERO.
300	--THREE HUNDRED.
450	--FOUR FIVE ZERO.
1,400	--ONE FOUR HUNDRED.
6,000	--SIX THOUSAND.
4,050	--FOUR ZERO FIVE ZERO.

63. LEADS. The lead element is included in the initial fire order only when firing at moving targets. Each 5-mil unit is called "one lead."

64. COMMAND TO OPEN FIRE. The command to open fire is FIRE. Strictly interpreted, however, it means FIRE WHEN READY. Since the command can either be given immediately or withheld, it constitutes a fire control element.

65. EXAMPLES OF INITIAL FIRE ORDERS.

a. GUNNER.
SHOT.
RIGHT FRONT.
TANK.
EIGHT HUNDRED.
ONE LEAD.
FIRE.

b. BOG.
CALIBER THIRTY.
FRONT.
DOUGHS.
THREE HUNDRED.
FIRE.

c. GUNNER.
HE.
DELAY.
TRAVERSE RIGHT.
STEADY. . . . ON.
ANTITANK.
ONE TWO HUNDRED.
FIRE.
TARGET? (by gunner).
ANTITANK.

d. NUMBERS TWO AND THREE.
WATCH MY BURST.
MY LEFT FRONT.
TRUCK.
ONE FIVE HUNDRED.
FIRE (to own gunner).
ON THE WAY.

e. GUNNER.
SHOT.
REFERENCE POINT.
RIGHT FOUR ZERO.
TANK.
ONE TWO HUNDRED.
FIRE.

f. GUNNER.
HE.
CONCRETE.
REFERENCE POINT.
LEFT ONE HUNDRED YARDS.
PILLBOX.
FOUR HUNDRED.
FIRE.

Section III. SENSINGS AND SUBSEQUENT FIRE ORDERS

66. ANNOUNCING SENSINGS. **a.** In direct fire, each round is sensed for deflection and range. Only the range sensings are announced.

b. Range sensings for HE and shot are discussed in the chapters allotted to conduct of fire with these types of ammunition.

67. SUBSEQUENT FIRE ORDERS. **a. Elements.** Elements of a subsequent fire order include a—

(1) Correction in deflection.

(2) Correction or change in range or elevation

(3) Command to fire.

b. Announcing subsequent orders. Subsequent fire orders for range or elevation are announced as corrections to the sight picture, handwheel setting, or elevation at which the last round was fired.

c. Corrections in deflection. (1) When firing against stationary targets, corrections in deflection are given as follows: RIGHT TWO; LEFT FIVE.

(2) When firing at moving targets, corrections in deflection are given as follows: ONE-HALF MORE; ONE LESS.

(3) Corrections in deflection are included only when there is a correction to be made.

d. Command to fire. See paragraph 64.

68. EXAMPLES OF RANGE SENSINGS AND SUBSEQUENT FIRE ORDERS.

a. SHORT.
RIGHT TWO.
UP TWO HUNDRED.
FIRE.

b. DOUBTFUL.
LEFT ONE ZERO.
FIRE.

c. OVER.
DOWN THREE HUNDRED.
FIRE.

d. TARGET.
FIRE.

e. TARGET.
ZERO LEADS.
FIRE.

f. SHORT.
UP TWO HUNDRED.
ONE-HALF MORE.
FIRE.

Section IV. CORRECTING AND REPEATING COMMANDS

69. REPEATING. a. If the loader or the gunner fails to understand any element of the fire order, he may request a repetition of that element by announcing the misunderstood element, using a rising inflection in his voice to denote a question.

b. When any crew member asks that the *deflection or range* element be *repeated*, misunderstanding is avoided by prefacing the *repeated* element with the phrase THE COMMAND WAS______. This phrase is used only when *repeating* a command previously given.

70. CORRECTIONS. **a.** In all direct fire orders, an incorrect command is *corrected* by saying CORRECTION, and giving the *correct* command. To correct a command for range, announce CORRECTION, RANGE_______.

b. If the command FIRE has been given, but not yet executed, the command SUSPEND FIRING is given. The corrected command is then given, followed by the command FIRE.

CHAPTER 9

LAYING WITH QUADRANT AND GRADUATED HANDWHEEL

Section I. GENERAL

71. USE. **a.** Frequently, it will be necessary to engage targets which are not easily identified by the gunner. In this case, the crew commander may—

(1) Lay the gun himself initially. Thereafter, the gunner lays on the target or on an auxiliary aiming point near the target. Adjustments in range and deflection are made on the sight.

(2) Order the piece to be laid for deflection by a shift from a reference point, using the azimuth indicator or panoramic sight. The piece is laid for elevation by use of the elevation quadrant or gunner's quadrant.

b. In addition to the instance outlined in a(1) and (2), the gun is laid for elevation with a quadrant in the following situations:

(1) When firing against targets without vertical profile.

(2) When it is desired to deliver a high volume of fire on a point target. (See sec. IV, ch. 14.)

Section II. GUNNER'S QUADRANT, M1

72. GENERAL. The gunner's quadrant, M1 (fig. 26), is used to—

a. Lay for elevation against point targets when

accuracy greater than that obtainable with the elevation quadrant, M9, is required.

b. Test and adjust the elevation quadrant, M9.

c. Measure the elevation or cant of the gun.

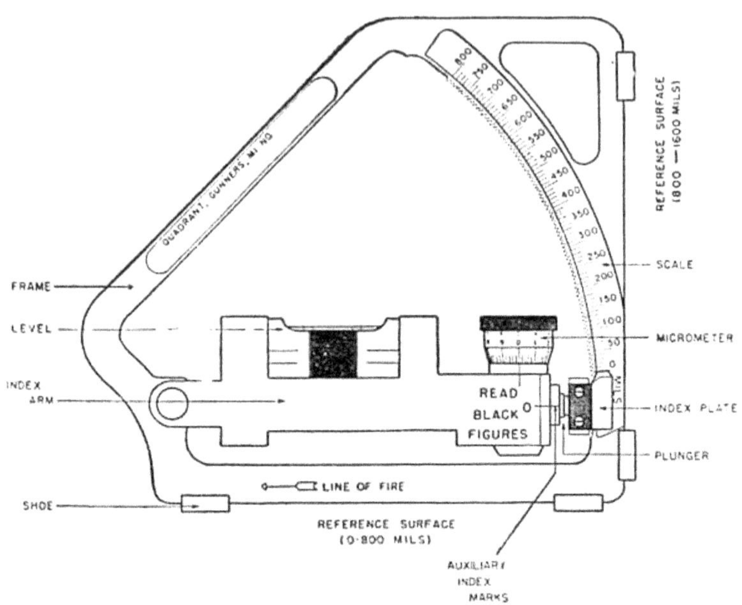

Figure 26. Gunner's quadrant, M1.

73. TEST OF ZERO SETTING (END FOR END TEST) Before use, the quadrant must be given the following tests:

a. The index arm and micrometer are both set at zero.

b. The quadrant is placed on the breech and the bubble centered by elevating or depressing the gun.

c. The quadrant is turned end for end on the breech. If the bubble recenters itself, the quadrant is in perfect adjustment at zero elevation.

If the bubble does not recenter itself, an attempt is made to center the bubble by turning the micrometer.

d. If the bubble can be recentered with the micrometer, the correction is *plus*. It is one-half the reading on the micrometer, and must be added to all settings.

e. If the bubble cannot be recentered by turning the micrometer, the index arm is set at minus 10 mils on the graduated arc (one notch below zero). Then the bubble is centered with the micrometer. In this instance, the correction is *minus* and must be subtracted from all subsequent readings. The correction is one-half the sum of the readings of the index arm and micrometer. Subtract the setting on the micrometer from 10 and divide the remainder by 2.

f. The error, as determined in a through e above, is set on the scales, the tube releveled and the quadrant tested again by turning end for end.

g. The quadrant may still be used by applying the corrections as indicated above. If the correction exceeds 0.4 mil, the quadrant should be sent to Ordnance for adjustment. Company personnel are not permitted to make any adjustment of quadrants.

74. TO LAY GUN. **a.** The index arm and micrometer are set at the required setting.

b. The quadrant is placed on the quadrant seats (leveling plates). If the breech has no quadrant seats, a level spot is selected on thc breechring and the quadrant is placed parallel to the bore.

The positions of the quadrant shoes are marked. The same positions are always used.

c. The quadrant must be facing in the correct direction, as indicated by the directional arrow on the same side as the scale used in making the setting. For *plus* quadrant elevations, the directional arrow points to the front; for a *minus*, to the rear.

d. By means of the elevating handwheel, the gun is elevated or depressed until the bubble is centered. The last motion of the elevating handwheel is always made in the same direction and against the greater resistance.

75. TO MEASURE ELEVATION. a. The micrometer is set at zero. The auxiliary indexes must be alined.

b. The quadrant is placed on the quadrant seats (leveling plates).

c. The quadrant must be facing in the correct direction as indicated by the directional arrow.

d. The index arm plunger is disengaged, and the arm lifted slowly until the bubble is approximately centered. The plunger is allowed to reengage the teeth of the frame.

e. The bubble is centered by turning the micrometer.

f. The arc and micrometer scale settings are read on the same side of the quadrant as the directional arrow used. In reading the micrometer, care must be taken to avoid 10 mil errors. If the auxiliary indexes are not alined when the micrometer reads "zero," the actual setting of the micrometer is 10 mils.

g. The elevation of the gun is the sum of the arc and micrometer settings. It is announced as QUADRANT______.

76. CARE AND MAINTENANCE. **a.** Care must be exercised to prevent unnecessary rough treatment that might cause the level vial to be broken, or the parts to be strained out of line.

b. The quadrant must be held securely when it is on the breech and must be removed from the breech before firing.

c. To prevent unnecessary wear and damage, the teeth of the plunger are kept clear of the teeth of the frame when moving the index arm.

d. The quadrant is kept in its case when not in use.

e. The level vial cover is kept closed when not in use.

f. The quadrant must not be left in the sun for prolonged periods.

g. Extreme care must be taken to prevent damage to the quadrant shoes.

h. Before the instruments are returned to their cases, they should be cleaned thoroughly with a dry, clean cloth. A thin film of light oil is applied to the shoe surfaces. One or two drops of oil are placed on the pivots and the plunger. Excess lubricant is wiped off to prevent accumulation of dust and grit.

Section III. ELEVATION QUADRANT, M9

77. GENERAL. The elevation quadrant, M9 (fig. 27) is used to lay the gun for elevation or to measure the quadrant elevation of the gun.

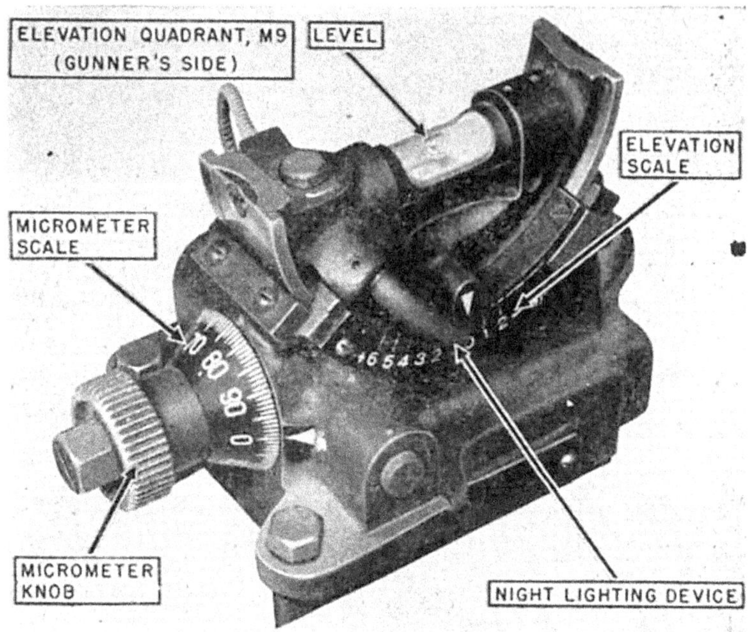

Figure 27. Elevation quadrant, M9.

78. TESTS AND ADJUSTMENT. a. An elevation of zero is set on an accurate gunner's quadrant, and the tube is leveled.

b. The elevation quadrant bubble is leveled by use of the micrometer knob. The elevation scales and micrometer scale on the gunner's side should read zero. The micrometer scale on the loader's side will read 50.

c. If the elevation scales do not read zero, the two screws on either side of both elevation scales are loosened. The zero graduations are moved until they are opposite the fixed indexes. The screws are then tightened and the scales checked.

d. If the micrometer scale does not read zero opposite the gunner's index, the bolt on the micrometer knob is loosened. Without moving the knob, the micrometer scale is rotated until the zero graduation is opposite the fixed index on the gunner's side. The bolt is tightened and the scale checked.

e. This completes adjustment of the elevation quadrant. No other adjustments are permitted by using troops.

79. TO LAY GUN. a. To set off an elevation of 135 mils, the micrometer knob is turned counterclockwise until the elevation index reaches the "1" on the plus side of the elevation scale. Turning of the micrometer knob is continued until a reading of "35" is opposite the micrometer index on the gunner's side of the quadrant.

b. The gun tube is elevated or depressed until the bubble is centered accurately. The last movement of the elevating handwheel is always made in the same direction and against the greater resistance.

c. To set off an elevation of *minus* 135 mils, the micrometer knob is turned clockwise until the index reaches the "1" on the minus side of the elevation scale. The tens and units, "35", are subtracted from 100 and the difference, "65" is set on the micrometer scale. For small changes, the micrometer knob is turned clockwise from a zero setting, the tens and units being counted as they pass the index.

80. TO MEASURE ELEVATION. **a.** The bubble is leveled with the micrometer knob.

b. The elevation of the gun is determined by reading the elevation and micrometer scales. If the elevation scale index is on the minus side of zero, the micrometer reading must be subtracted from 100 to obtain the tens and units of the minus elevation. (See par. 79c.)

81. CARE AND MAINTENANCE. **a.** In addition to the usual care of sighting equipment, the following steps to protect the quadrant should be taken:

b. The elevation quadrant should be kept covered when not in use to prevent dust and grit from collecting on the locating surfaces.

c. The cover on the level vial should be closed when not in use.

d. The felt pads at the ends of the segment should be kept pliable by occasionally applying a few drops of oil.

Section IV. LAYING

82. INITIAL LAYING WITH AZIMUTH INDICATOR AND QUADRANT. **a.** In laying for deflection, the gunner must take up lost motion in the traversing mechanism. This is done by making the final motion in the direction in which it is more difficult to traverse. If there is no direction of greater resistance, the gunner always makes the final movement for each setting in the same direction. It is often necessary to traverse beyond the target and then come back on it.

b. The gunner zeros his azimuth indicator or selects an aiming point for his telescopic sight. The range and the angle of site, if any, are announced by the crew commander. For example: Firing at a target at a range of 1,200 yards and with the angle of site estimated as plus 5 mils, the crew commander announces ONE TWO HUNDRED, UP FIVE. The gunner refers to the firing table and gets the elevation for 1,200.

(1) When using the graduated handwheel (fig. 28), the firing table elevation is set on the quadrant, and the bubble is leveled. The angle of site is then set off in clicks on the graduated handwheel.

Figure 28. Graduated elevating handwheel.

(2) If the gun is not provided with the graduated handwheel, the quadrant elevation is computed by applying a plus or minus angle of site to the range elevation obtained from the firing table. The quadrant elevation is then set on the quadrant, and the bubble is leveled.

c. In laying for elevation, the gunner must be certain to take up lost motion in the elevating mechanism. The procedure is similar to that given for eliminating lost motion from the traversing mechanism.

d. The gunner must verify the laying for direction and elevation after the breech is closed.

83. SUBSEQUENT LAYING WITH AZIMUTH INDICATOR AND QUADRANT. **a.** Corrections in deflection are announced in mils and applied to the azimuth indicator.

b. Corrections in elevation are announced in mils. When using the graduated handwheel, corrections are applied directly to the elevation of the tube by moving the handwheel one click in the proper direction for each mil. If there is no graduated handwheel, the announced changes are made in the setting on the quadrant, and the bubble is releveled.

c. As in initial laying, the gunner must take up lost motion in the elevating and traversing mechanisms and must verify the laying after the breech is closed.

d. In using the graduated handwheel, the gunner must grip the wheel throughout the adjustment. If he does not, the wheel turns, and the accuracy of the adjustment is lost.

e. Even when using the graduated handwheel and azimuth indicator for range and deflection, the gunner makes every effort to obtain a distinct sight picture on every round. This allows him to observe each round, and possibly, to identify a target which he may not have seen initially.

CHAPTER 10

CONDUCT OF FIRE—GENERAL

84. DEFINITION. Conduct of fire is the technique of placing fire upon a target. In general, the technique employed will depend upon the type of ammunition used.

85. SHOT. a. At normal direct fire ranges, the tracer element of shot allows the crew commander to determine accurately the distance (horizontal or vertical) between the center of the target and that point in the trajectory at which the projectile is closest to the target. Subsequent fire orders are based on data which are carefully computed to move the strike to the center of the target. The technique is limited in application by the tracer burn-out point. When firing shot, the second round should always be a hit.

b. When firing high velocity guns, it is often impossible for the crew commanders to observe the tracer from the vehicle due to obscuration caused by blast, dust, smoke, and heat waves. When such situations arise, the crew commander should observe the fire from the flanks or have an adjacent vehicle accomplish the sensing and adjusting. If such methods are impractical, the crew commander will have to resort to methods prescribed in paragraphs 102 and 107 in sensing and adjusting his fire. Information gained from previous firing and knowledge of terrain should assist in sensing if tracer and point of impact cannot be observed.

86. HIGH EXPLOSIVE AND WHITE PHOSPHORUS SMOKE. **a.** HE ammunition contains no tracer element. The procedure is to inclose the target in a range bracket. The size of the bracket is split successively until effect is obtained on the target.

b. The adjustment of white phosphorus smoke is similar to the procedure used with HE.

CHAPTER 11

CONDUCT OF FIRE—SHOT

Section I. GENERAL

87. VERTICAL PROFILE. Shot is used primarily against targets which have a vertical profile. Conduct of fire with shot is adjustment within a vertical plane. The purpose of firing shot is to obtain penetration.

88. TRAJECTORY. To adjust fire correctly with shot and to utilize fully the capabilities of the cannon, the crew commander and gunner must understand the trajectory.

Section II. TRAJECTORY

89. ELEMENTS OF TRAJECTORY. a. Effect of deflection errors on strike. The path of the projectile may be considered as a rigid curve, pivoted at the gun and capable of being raised or lowered about its origin through small vertical angles without materially changing the shape of the curve. (See figs. 29 and 30.) The trajectories of all high velocity guns are relatively flat. For this reason, when firing at targets with vertical profile, a round which passes a target at the correct elevation, but at the wrong deflection, may travel far beyond the target before it strikes the ground.

b. Effect of range on strike. Figure 31A shows graphically the paths of three rounds fired at a tank at a range of 500 yards; one passing the tank at a point 1 foot below the top of the turret, one at the vertical center of mass, and one at a point 1 foot above the ground. Figure 31B, C, and D,

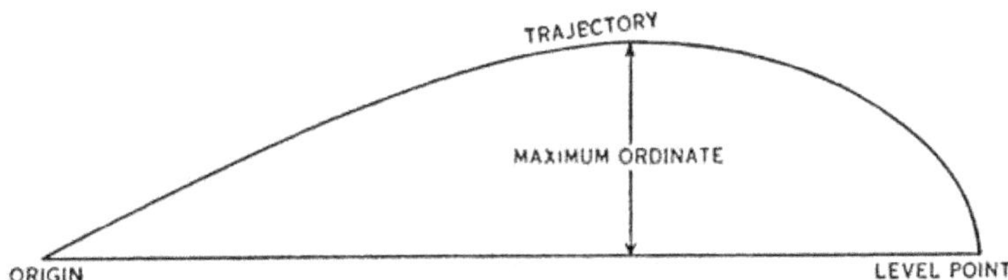

Figure 29. Trajectory.

show the trajectories of similar rounds fired at a tank at ranges of 1,000, 1,500, and 2,000 yards, respectively. The projectile in every instance, if it missed the tank in deflection, would travel far

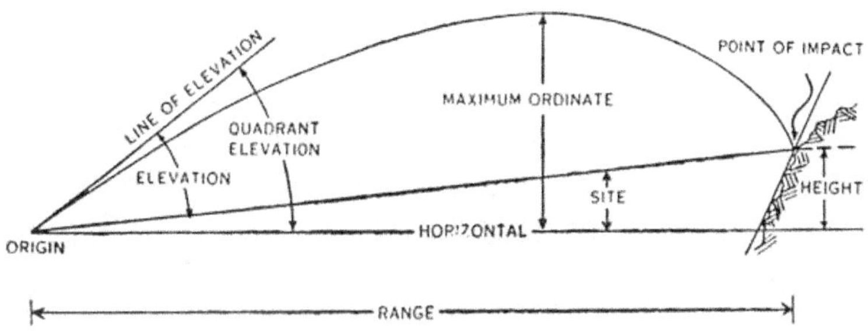

Figure 30. Elements of trajectory, showing point of impact on raised ground.

beyond the target before it struck the ground. Note that, at the range of 500 yards, the center round travels more than 400 yards beyond the tank before striking the ground.

90. SUMMARY. From a study of figure 31 two things are obvious.

a. First. At the range at which direct fire usually is employed, every round fired with correct range against a vertical target, such as a tank.

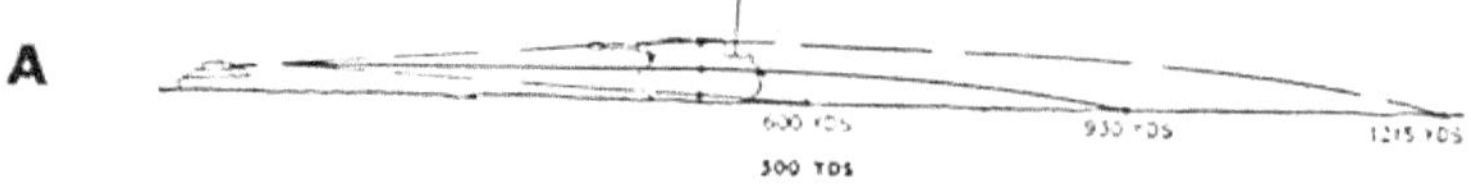

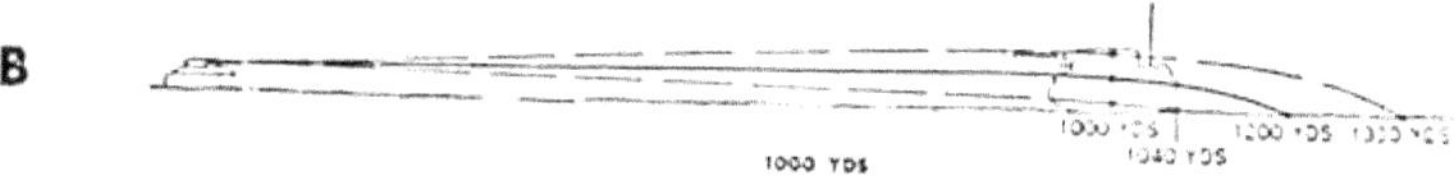

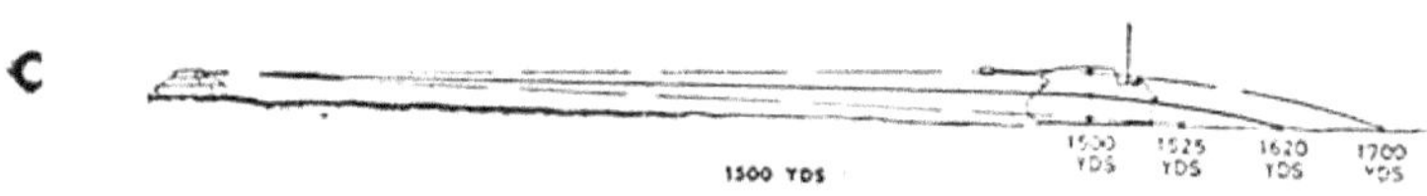

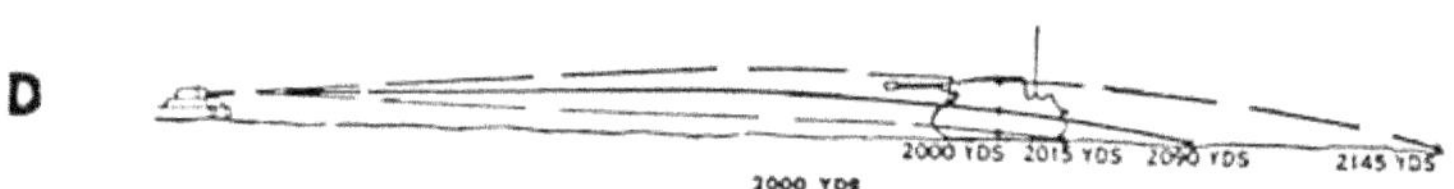

Figure 31. Effect of range on strike of projectile when target is at 500, 1,000, 1,500, and 2,000 yards.

will (if it goes *past* the target) strike the ground a considerable distance beyond. Every hit on the cloth targets used in training will strike the ground well beyond the target. This knowledge is important in eliminating the common error of sensing a round "over" merely because it strikes the ground *beyond* the target. Actually, the *range* may have been correct.

b. Second. On level or uniformly sloping ground, the distance which a round fired at the correct range strikes beyond the target decreases as the range to the target increases. Thus, to adjust solely by judging the distance a round strikes beyond the target, presents serious difficulties even on level ground because of the variation of this distance with the range.

91. EFFECT OF TERRAIN. Compare figure 32 with figure 31A. The ranges to the targets are the same in both instances. However, in figure 32, terrain variations change the strike of the projectile.

92. RESULTS. Two facts are thus determined.

a. Every round fired with range correct will strike some distance beyond the target if it passes or goes through the target.

b. The distance which the round strikes beyond the target varies so greatly with the range and with the shape of the terrain that adjustments cannot be based upon an estimate of this distance.

93. VERTICAL TARGETS. The trajectory and the

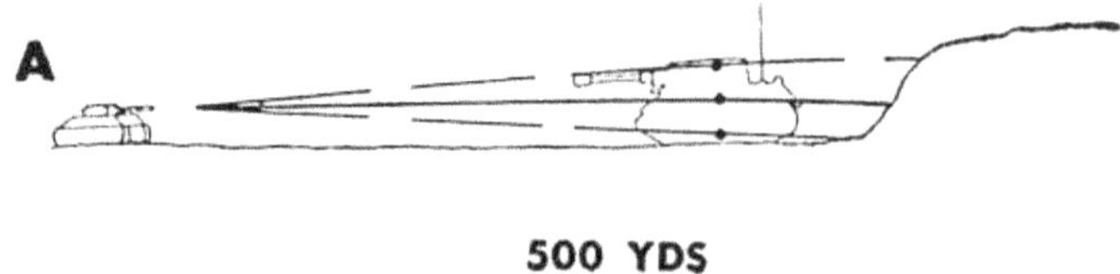

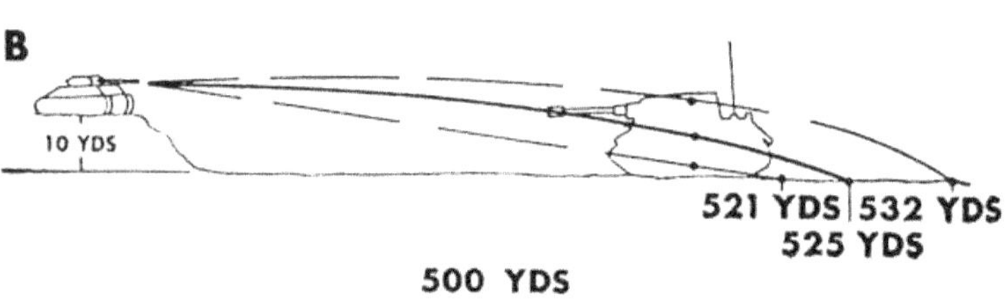

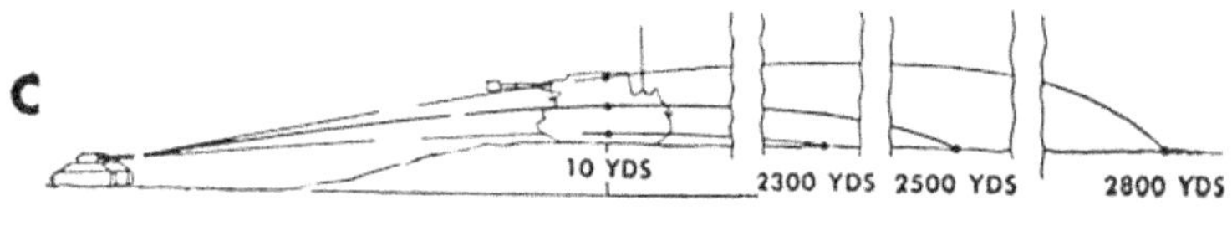

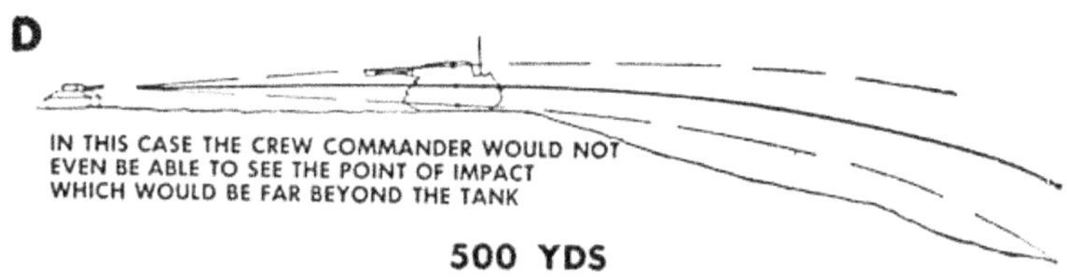

Figure 32. Effect of terrain on strike of projectile.

problem of sensing may be considered from another angle. When a rifleman, firing on a target range, gets a shot on the target 2 feet above the center of the bull's-eye, he must know the amount of range change to make on his sight to put the next round in the center of the black. When firing high velocity cannon, the problem is similar to that of the rifleman. The rifleman shoots at a vertical target. Likewise, most of the targets engaged with shot extend above ground and may be considered as vertical.

94. VERTICAL DISPLACEMENT. a. The 76-mm gun fired at a target at 500 yards with range 500 will hit at *A* (the vertical center of mass of the target). (See fig. 33.)

b. Using the same aiming point, with range 600, the round will strike at *B*, approximately 1 foot above *A*. With range 400, the projectile will strike at *C*, approximately 1 foot below *A*.

c. Thus, for every 100-yard range change on the sight reticle, with the target at 500 yards, the point of strike is moved approximately 1 foot vertically on the target.

d. Similarly, figures 34 to 36 inclusive show the change in strike for each 100-yard range change on the sight with the target at 1,000, 1,500, and 2,000 yards.

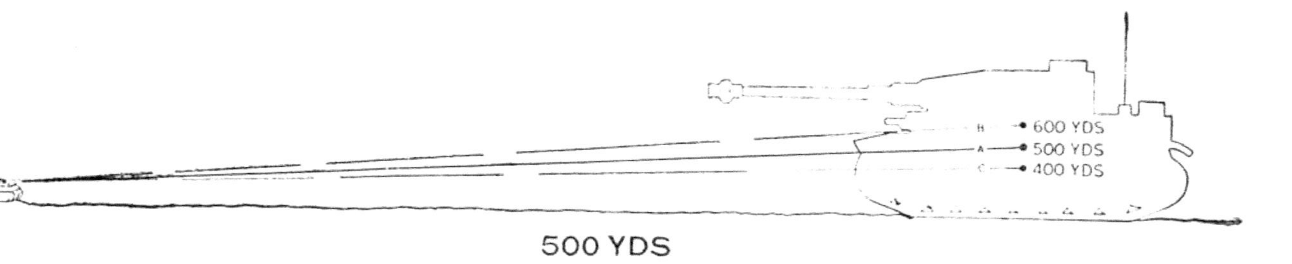

Figure 33. Vertical displacement at 500 yards.

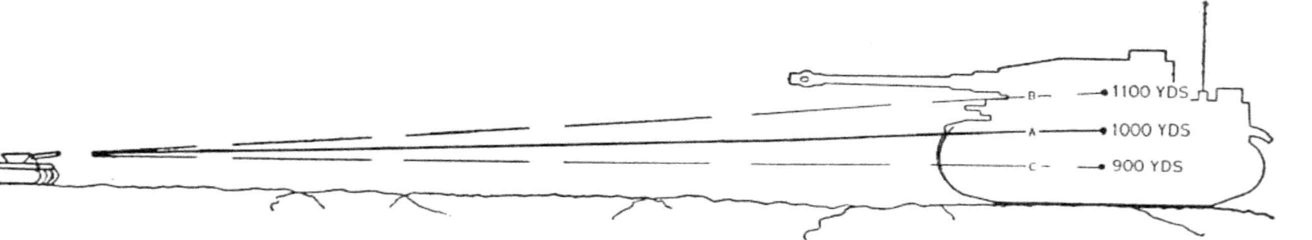

Figure 34. Vertical displacement at 1,000 yards.

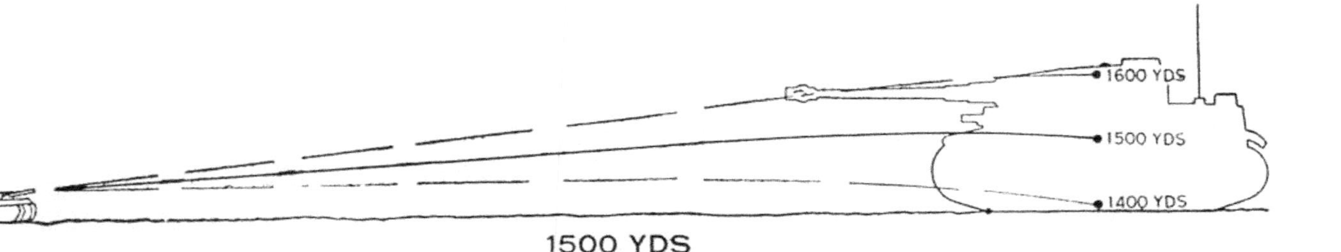

Figure 35. Vertical displacement at 1,500 yards.

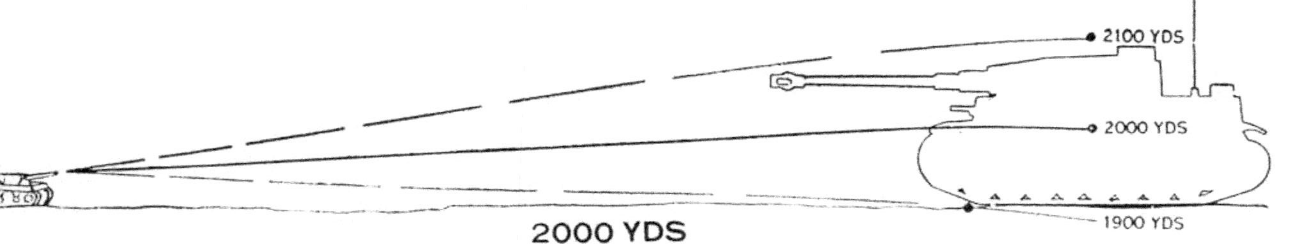

Figure 36. Vertical displacement at 2,000 yards.

e. Approximate vertical change in feet for 100-yard range change.

Weapon—ammunition	Target range—yards			
	500	1,000	1,500	2,000
37-mm, M6:				
HE M63	1	3	6	10
APC M51	1	2	4	5
AP M74	1	3	7	12
75-mm Gun, M3 and M6:				
HE M48 s/c	2	5	7	11
HE M48 n/c	4	8	13	19
APC M61	2	4	7	11
AP M72	2	5	9	13
76-mm Gun, M1A2:				
HE M42A1	1	2	4	6
APC M62	1	2	4	6
AP M79	1	3	5	8
90-mm Gun, M3:				
HE M71	1	2	4	5
APC M82	1	2	4	5
AP M77	1	3	5	8
105-mm Howitzer, M4:				
HE M1 c/7	3	7	13	18
HE M1 c/6	5	11	18	25
HEAT M67	5	12	20	28

95. CONCLUSION. The exact distance which a range change of 100 yards will move the point of strike on a vertical target is not important. It is important, however, for the crew commander and gunner to realize the possibilities of moving the strike up and down the face of a vertical target by changing the range. It is also important that the crew commander take the small displacement at shorter ranges into account when computing his subsequent fire order. When firing at short range targets which have vertical profile, relatively large changes in range can be made without causing successive rounds to miss the target.

96. DISPERSION. **a.** If 100 rounds were fired from a gun laid with the correct range on the center of a vertical target, they would not all strike the center. They would be dispersed in a symmetrical pattern around the center (center of impact). If the total vertical distance between the top and bottom of the shot group is divided into eight equal parts, it will be found that the shots are dispersed in the percentages shown in figure 37.

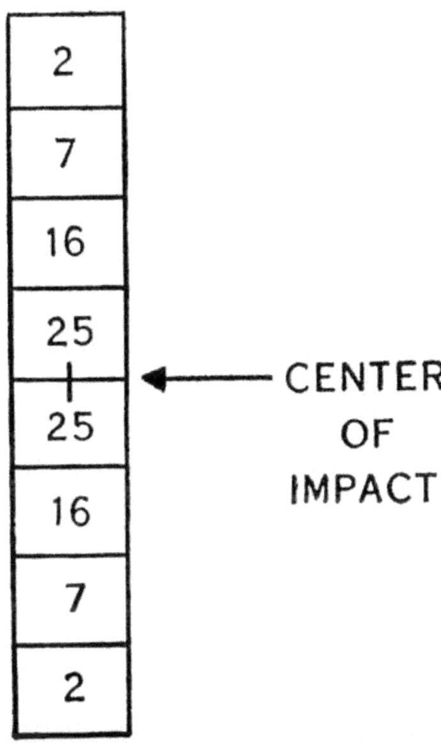

Figure 37. Dispersion pattern.

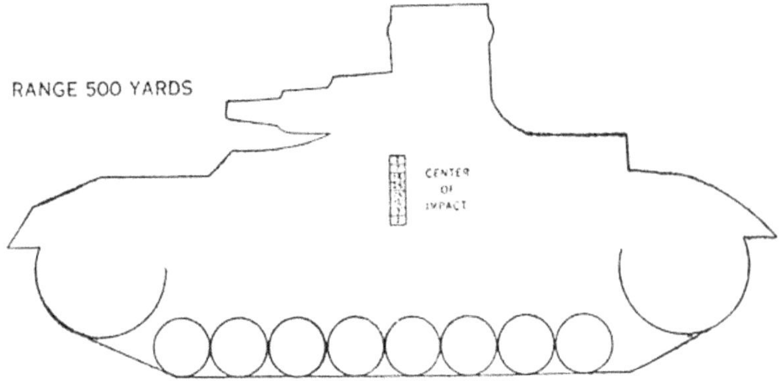

Figure 38. Dispersion at 500 yards.

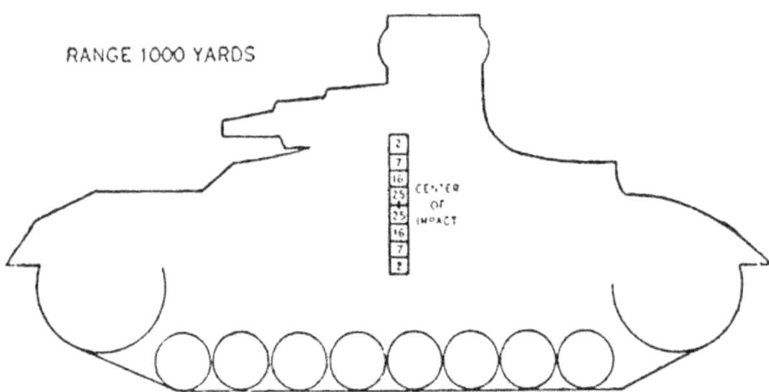

Figure 39. Dispersion at 1,000 yards.

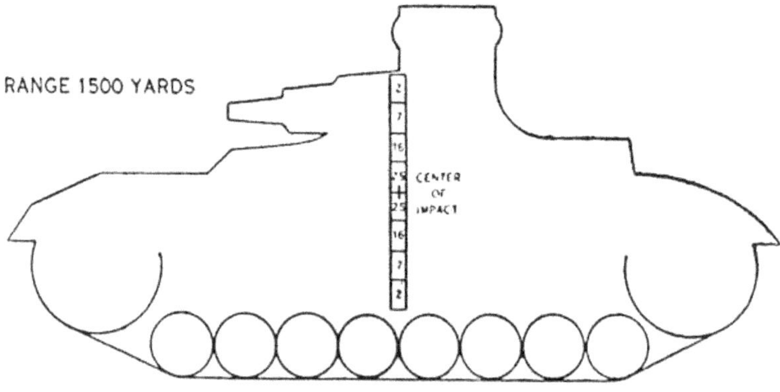

Figure 40. Dispersion at 1,500 yards.

b. At short ranges, the vertical dispersion of high velocity cannon is extremely small. Figure 38 illustrates the dispersion pattern of the 76-mm gun firing at a tank 8 feet high at a range of 500 yards. Figures 39, 40, and 41 illustrate the increase in size of the dispersion pattern at greater ranges. It is quite evident that the effects of dispersion need not be considered when firing at tanks at normal direct fire ranges. In each instance, however, it has been considered that the entire tank was in view. Figures 42 and 43 illustrate the effects of dispersion when firing against a tank in partial defilade. Similar results can be expected in firing at other targets with relatively little profile.

c. The effects of dispersion do not directly affect conduct of fire procedure. Crew commanders must realize, however, when firing at long direct fire ranges or at targets of little profile, that dispersion is a factor which must be considered. An appreciation of the effects of dispersion is part of a comprehensive preparation for conduct of fire.

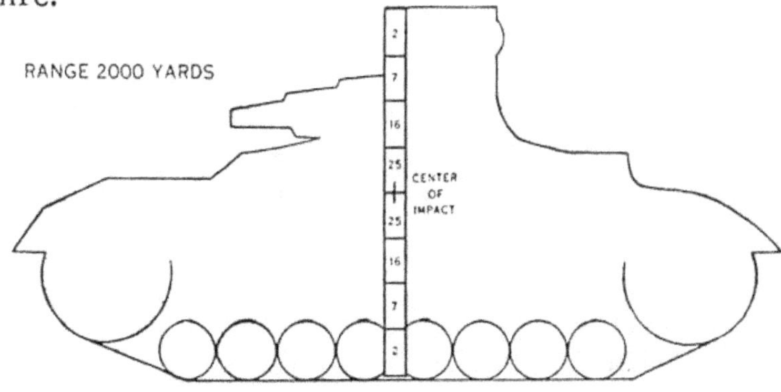

Figure 41. Dispersion at 2,000 yards.

Section III. SENSING AND ADJUSTMENT

97. SENSING SHOT. A sensing for shot involves a determination of whether the projectile—

a. Strikes or penetrates the target—sensed as "target."

b. Passes over the target—sensed as "over."

c. Strikes the ground between the gun and the target—sensed as "short."

d. Passes the target so that a range sensing cannot be made because of the deflection error—sensed as "doubtful."

98. POINT OF STRIKE AND TRACER. **a.** When shot strikes armor plate, a red flash is produced, and sensing is easy. Against pillboxes or other types of targets, particles of concrete or other material may be seen flying from the target.

b. If the round passes the target and is not a hit, or if in practice firing it passes through a cloth target panel, the path of the tracer furnishes more accurate information than does the point of strike on the ground. (See par. 89.)

99. MEASURING TARGET. In all conduct of fire against targets with vertical profile, the first step is to measure the height of the target in mils. The measurement is taken to the nearest whole mil. It can be made with any fire control instrument containing a mil scale.

100. SENSING "OVERS". When the round passes over the target, the problem is to determine the

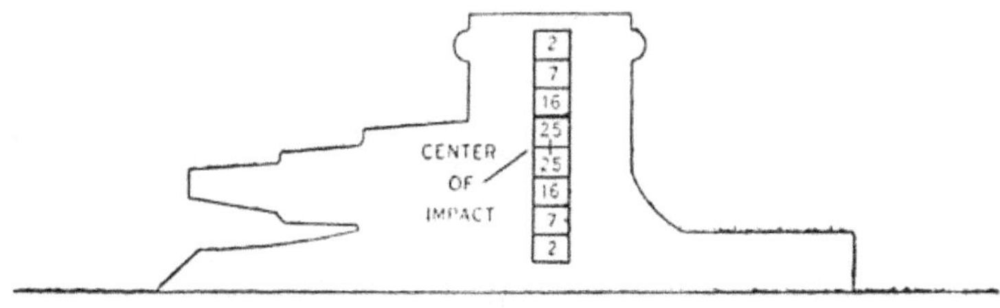

Figure 42. Dispersion at 1,000 yards against a tank in partial defilade.

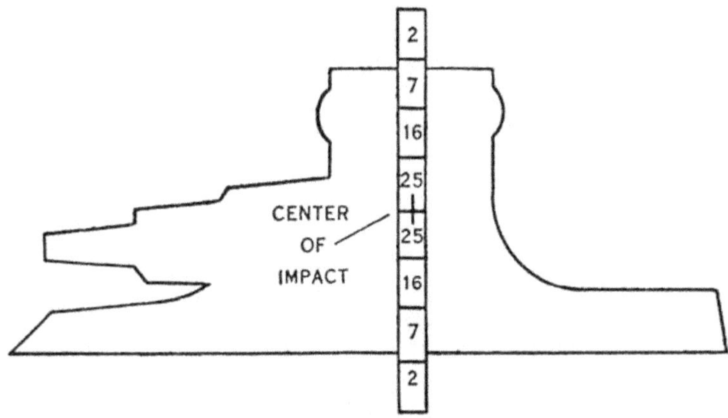

RANGE 1500 YARDS

Figure 43. Dispersion at 1,500 yards against a tank in partial defilade.

angular distance between the projectile and the center of the target. *Observe the tracer just as the projectile passes the target.* The eyes are focused on the target, and, as the tracer passes, its height is noted with respect to the center of the target. Sensing in this manner is difficult and must be stressed during subcaliber firing. The tendency to sense the projectile at the maximum ordinate, rather than as it passes the target, must be corrected. There are two methods of determining the amount of angular displacement. The method to be used depends upon the range to the target, the position of the crew commander and the amount of obscuration caused by flash smoke, dust, and heat waves.

a. The displacement between the projectile and the target is estimated, using the mil height of the target as a scale or yardstick. (See fig. 44①.)

b. The displacement is measured, using the mil scale in field glasses. (See fig. 44②.)

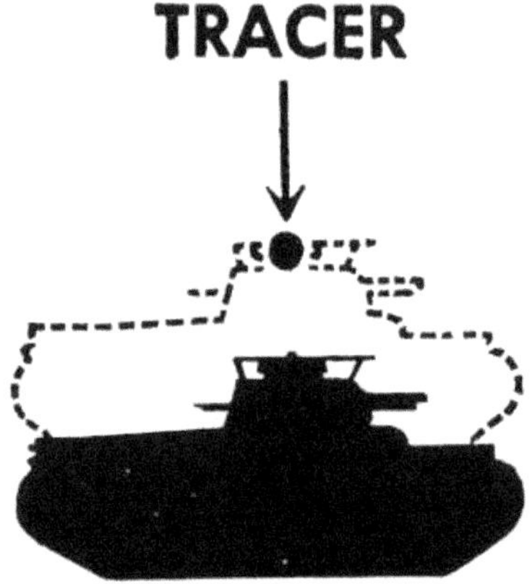

① *Determining vertical displacement by using the mil height of the tank as a yardstick.*

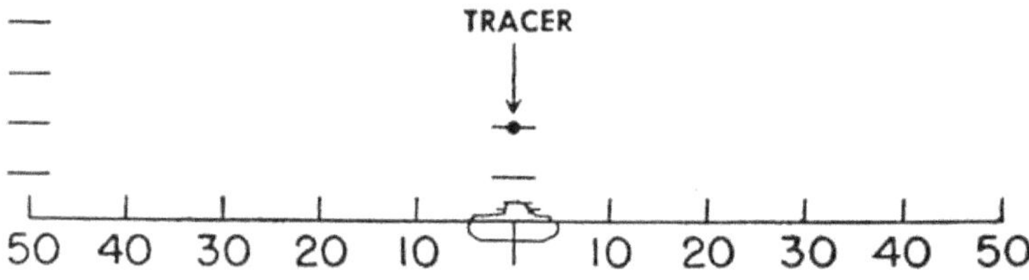

② *Measuring the vertical displacement with field glasses*
Figure 44.

101. ADJUSTING "OVERS". **a.** The vertical displacement in mils between the projectile and the center of the target is converted into "c's" (a "c" is the change in elevation to effect 100-yard change in range) and is expressed as a range change in hundreds of yards. A decrease of 100 yards in range is made for each "c" of displacement.

b. For example: 76-mm APC is fired at a tank at an estimated range of 800 yards. The tank has been measured to be 4 mils high. The first round appears a full tank height over the center of the tank as the tracer passes. The verticle angular displacement, therefore, is 4 mils. Since "c" at this range for the 76-mm gun is "one" (fig. 45), the subsequent fire order is DOWN FOUR HUNDRED.

102. SENSING AND ADJUSTING "SHORTS". When the projectile strikes the ground between the gun and the target, the procedure of the crew commander is as follows:

APPROXIMATE VALUES OF "c"

Weapon	Range in yards	Approximate values of "c"
37-mm Gun, M6	0 – 1,500	1
57-mm Gun, M1	0 – 3,000	1
75-mm Guns, M3 and M6	0 – 4,500	2
76-mm Gun, M1	0 – 4,000	1
90-mm Gun, M3	0 – 5,000	1
105-mm Howitzer, M4,	0 – 2,000	2
w/chg VII	2,000 – 5,000	4

For greater ranges use firing tables.

Figure 45. Change in elevation for 100-yard range change.

a. When the distance that the projectile strike short can be estimated, the subsequent fire order must compensate in both horizontal and vertica planes.

(1) If the range is merely increased by the amount that the round was short, the next round would strike at the base of the target. (See fig 46.) An additional increase in range must be made to raise the strike to the center of the target. The latter amount may be determined by taking one-half of the target height in mils, converting it into "c's" (fig. 45) and expressing it in hundreds of yards. The range is increased 100 yards for each time that one-half the target height is divisible by "c." The computed range increase is added to the amount that the round landed short of the target.

(2) For example: 76-mm APC is fired at a tank at an estimated range of 800 yards. The tank has been measured as 4 mils high. The round strikes an estimated 200 yards short of the target. The range must be increased 200 yards to get to the base of the tank. One-half the tank height is 2 mils, and "c" for the 76-mm gun is 1 mil. Therefore, the range must be increased an additional 200 yards to hit the center of mass. The subsequent fire order is UP FOUR HUNDRED.

b. When the distance that the round strikes short of the target is not known, or cannot be estimated, an arbitrary range change is made.

(1) For high velocity guns, the initial change should be at least 400 yards. Larger changes should be made whenever the situation warrants.

The change should always be large enough so that the next round will hit or bracket the target.

(2) The same reasoning governs the size of the initial range change when firing medium velocity guns. The second round should always hit or bracket the target. Because of the nature of the trajectory, an initial change of only 200 yards may be justified when firing at ranges of 1,500 yards or less.

103. COORDINATION. a. *When the crew commander makes a definite range sensing,* that is, OVER, SHORT, or TARGET, *the gunner follows the range change in the crew commander's subsequent fire order.* The crew commander must, however, be sure that he is correct before making a positive sensing.

b. When the crew commander's sensing is LOST or DOUBTFUL, but the gunner has made a definite sensing, the gunner may make a range change in accordance with his own sensing. The gunner announces his sensing and range change to the crew commander. For example: Crew commander, "DOUBTFUL, FIRE"; gunner, "SHORT, UP FOUR HUNDRED, ON THE WAY."

c. In tank destroyer gunnery, wherein the crew commander is usually dismounted from the turret, the gunner will make changes only on the command of the crew commander.

104. SENSING AND ADJUSTMENT FOR DEFLECTION.
a. Stationary targets. Deflection errors can always

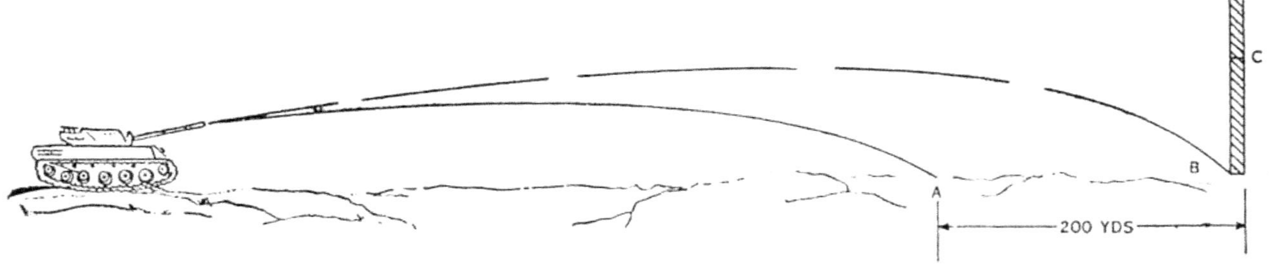

Figure 46. Adjustment of fire upon a vertical target.

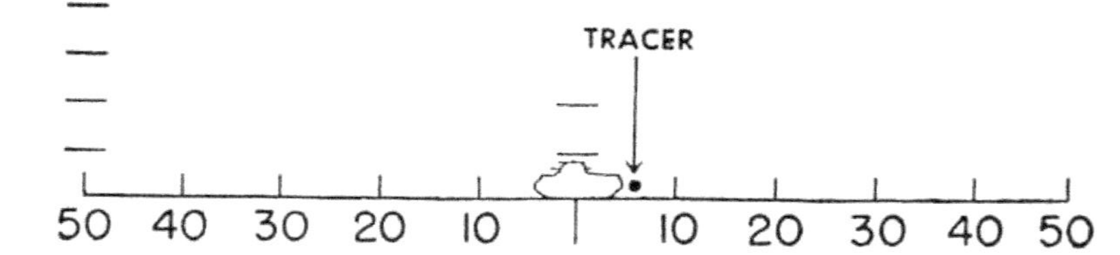

Figure 47. Measuring deflection error

be measured exactly. There is no justification for a deflection error after the first round.

(1) The error may be measured, using field glasses. (See fig. 47.)

(2) The error may be estimated, using the mil width of the target as a yardstick. (See fig. 48.)

(3) Corrections are announced in mils, as LEFT THREE, or RIGHT TWO.

b. Moving targets. Deflection errors may be difficult to sense when the projectile strikes the ground short of the target. In such instances, the crew commander must use his best judgment in sensing for deflection. Positive deflection sensings can be made when range is approximately correct, or over. *Corrections are announced in leads.* If the projectile is sensed as 5 mils from the center of the target, the command is ONE MORE (LESS). If the error is approximately 2 or 3 mils, the command is HALF MORE (LESS). Lead changes of less than one-half lead are impracticable, and are not made. For example: The target has been measured to be 5 mils wide. As the projectile passes the target, it appears to be a full target width in rear of the center of the target. The command is ONE MORE.

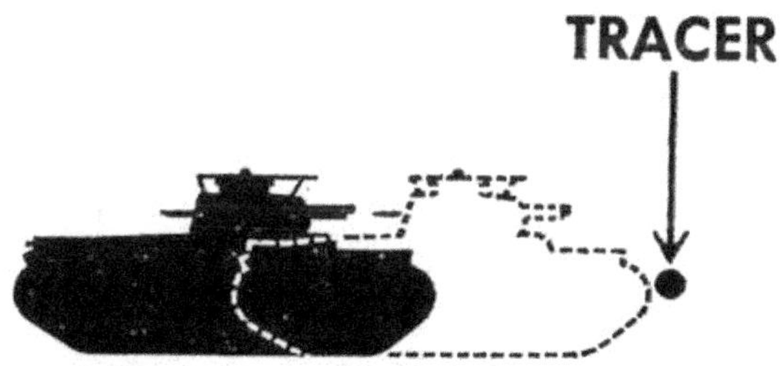

Figure 48. Estimating the deflection error by using the mil width of the tank as a yardstick.

CHAPTER 12

CONDUCT OF FIRE—HE

105. GENERAL. **a.** High explosive is employed primarily against targets without vertical profile and against targets which cannot be as effectively engaged with a machine gun.

b. The determination of the relationship between the point of an HE burst and the target is a sensing. The observer bases his sensings on what he sees while the burst is before his eyes. He cannot rely on his memory, but must make his sensing instantaneously. The blast of the bursting charge facilitates sensing.

c. Because of the difficulty in determining when fire is effective, the procedure followed is to inclose the target in a bracket. (See fig. 49.) A target is bracketed for range when successive rounds at the same or different elevations have produced an "over" and a "short." At direct fire ranges, normal range dispersion in medium and high velocity guns is not likely to produce an "over" and a "short" at the same elevation.

106. SENSING HE. **a.** If the target is clearly defined (silhouetted) against the burst, the range is "over."

b. If the target is obscured by the burst, the range is "short."

c. If a definite range sensing cannot be made because of a deflection error, the round is sensed as "doubtful." Definite sensings are made only

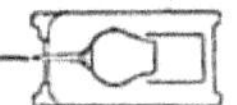

Figure 49. "Overs" and "shorts."

when their accuracy is certain. If a round is "doubtful," the procedure is to correct the deflection and obtain a line shot, so that the range sensing on the next round can be positive.

d. Positive sensings can be made when the burst is off the line, if the terrain reveals clearly the location of the burst in relation to the target.

e. The range is correct, and "target" may be sensed—

(1) When the target is seen to change shape or part of it is blown into the air.

(2) When the visible part of the target is seen to be inclosed by the pattern of striking fragments.

f. The range is correct when an "over" and a "short" have been sensed at the same range or elevation setting, or when the bracket has been split to the smallest practicable elevation change.

g. If the target is obscured and then immediately silhouetted by smoke or dust, or vice versa, the range is nearly that of the target.

h. Rounds which are not seen are sensed as "lost." Knowledge of the terrain may justify sensings of "lost over" or "lost short" when the deflection is known to be approximately correct. A bold range change is made to bring the next round into view.

i. For ricochet fire, sensing is on the smoke ball, the flash, and the fragments striking the ground, not on the dust thrown up by the strike of the projectile.

107. ADJUSTMENT OF HE. a. When the target is on the forward slope of a hill or has vertical profile, the procedure may follow that outlined in section III, chapter 11.

b. At all other times, an arbitrary change in range is made.

(1) For high velocity guns, the initial change should be at least 400 yards. Larger changes should be made whenever the situation warrants. The change should always be large enough so that the next round will hit or bracket the target.

(2) The same reasoning governs the size of the initial range change when firing medium velocity guns. The second round should always hit or bracket the target. Because of the nature of the trajectory, an initial change of only 200 yards may be justified when firing at ranges of 1,500 yards or less.

c. In firing at targets with vertical profile, dispersion does not directly affect conduct of fire. Horizontal dispersion follows the same pattern as does vertical dispersion (fig. 37), but is considerably larger. The dispersion pattern of the 76-mm gun is approximately 100 yards long at direct fire ranges. This factor must be considered in conduct of fire against targets without vertical profile. (See par. 108d.)

d. Once the target has been bracketed, the size of the bracket is split successively until—

(1) Effect is secured on the target; or

(2) The gunner has made the smallest elevation change practicable when the dispersion of the gun is taken into consideration. (See par. 108.)

108. FIRE FOR EFFECT. a. To obtain a direct hit, or to take advantage of the maximum depth of the burst pattern (fig. 50), correct deflection is essential.

b. *When a target hit is positively observed,* one or more rounds should be fired with the same laying to insure destruction.

c. *With medium velocity guns,* adjustment continues until effect is secured on the target. Dispersion is small, and the gunner can make small range changes on his reticle or graduated handwheel.

d. *With high velocity guns,* dispersion makes small changes in range or elevation unnecessary. When the bracket has been split to the smallest practicable elevation change (50 yards), rounds are fired at that range until effect is secured on the target. In this instance, dispersion will favor the gunner and will result in a hit or effect on the target.

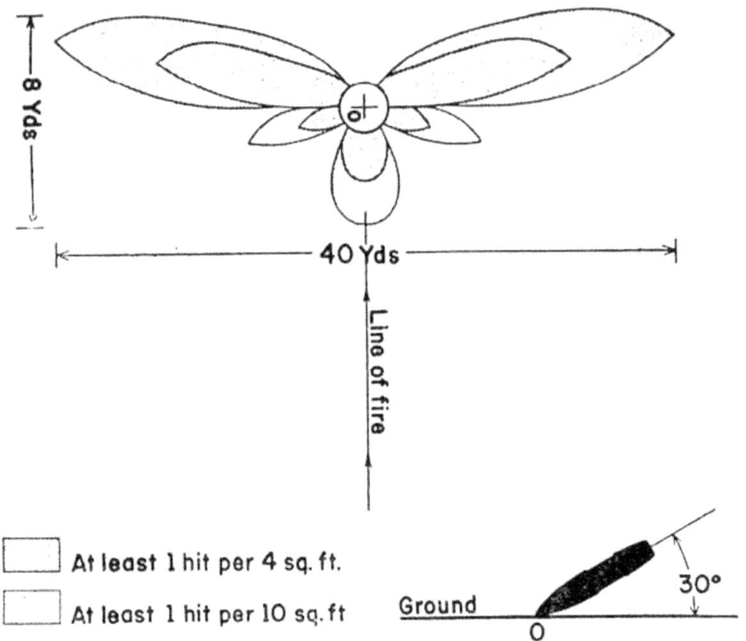

Figure 50. Impact burst pattern of 75-mm Shell HE M48, with superquick fuze.

109. RICOCHET FIRE. **a.** Against personnel in the open or in shallow trenches ricochet effect should be sought. For characteristics of ricochet action see FM 6–40.

b. Ricochet fire has several advantages. In figure 51, *XYZ* is the path of the ricocheting projectile. A round with a superquick fuze setting would burst at *Y*. In the latter case (superquick), there would be a direct hit on an antitank gun shield, if the gun were located anywhere within space *A*. If the fuze were set at delay, however, a direct hit would be obtained if the antitank gun were located anywhere within the entire space *B*. The burst of the projectile with delay fuze occurs at *Z*.

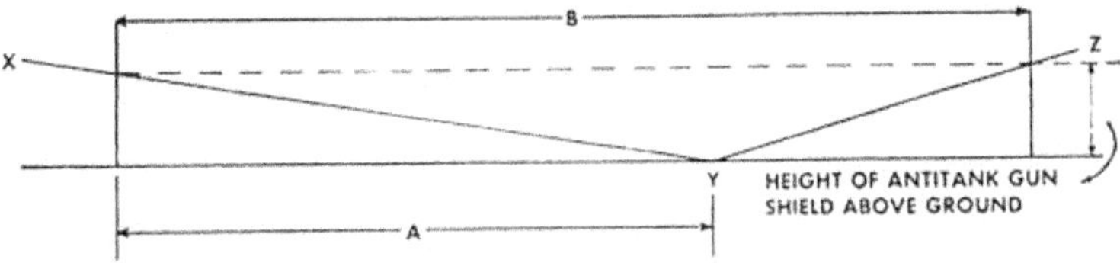

Figure 51. Increase in danger area by use of ricochet.

CHAPTER 13

GUNNERY TRAINING

Section I. GENERAL

110. TEAMWORK. Success in battle depends upon close coordination within the crew. Each man must be thoroughly acquainted with his own functions as well as with the duties of every other member of the crew. Conduct of fire, however, depends primarily upon the skill and coordination of the gunner and the crew commander. The training of these two members of the crew must be integrated carefully. The training of each will be treated as a separate subject. The gunner-crew commander team training starts with 1,000-inch firing and is perfected in field firing.

Section II. GUNNER

111. INTRODUCTION. The first phase of gunner training is instruction in sights and sight adjustment. The gunner should be given ample practice in setting sight pictures on the gunner trainer.

112. GUNNER TRAINER. a. Description. The trainer consists of two parts:

(1) The movable, transparent paddle, containing a sight reticle and a set of cross hairs to represent the muzzle bore sight. (See fig. 52.)

(2) The target board, consisting of a solid background with a tank silhouette and an object

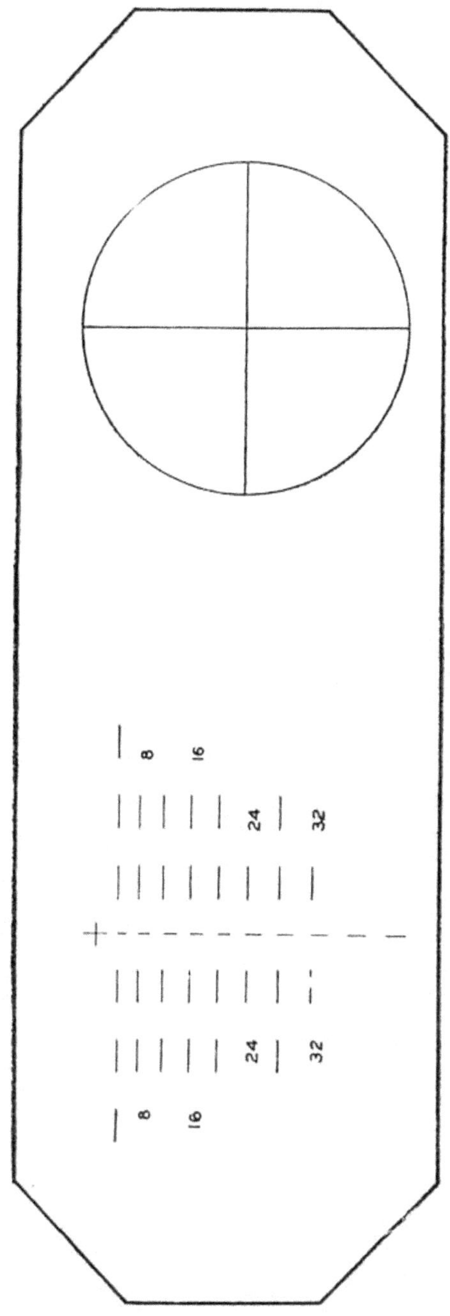

Figure 52. Transparent paddle.

suitable as a distant aiming point for boresighting. (See fig. 53.)

b. Construction. The trainer may be prepared locally in the following manner:

(1) *Movable paddle.* Exposed, bleached X-ray film; parts of old car curtains; or any other pliable, transparent material may be used. The material is cut to convenient size. On it is painted or drawn the reticle and a circle with cross hairs to represent the muzzle prepared for boresighting. (See fig. 52.) India ink or any good black paint is satisfactory.

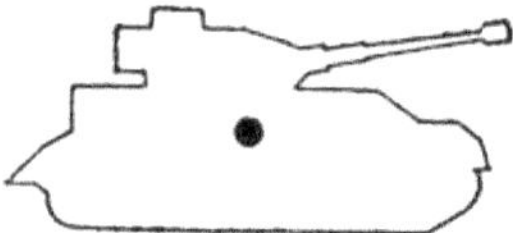

Figure 53. Target board.

(2) *Target board.* The outline of a tank, to be used as a target, and of a house, or some other object suitable for use as a distant aiming point for boresighting, are drawn on a light-colored, solid background. (See fig. 53.)

(3) *Size.* The size of the trainer should be that best suited to the type of instruction to be given. It may be made small enough to fit in a shirt pocket, or large enough for use with a large group.

c. Use. This simple training aid is invaluable for the critique of firing exercises. The instructor must insist on pin-point accuracy in setting off sight pictures.

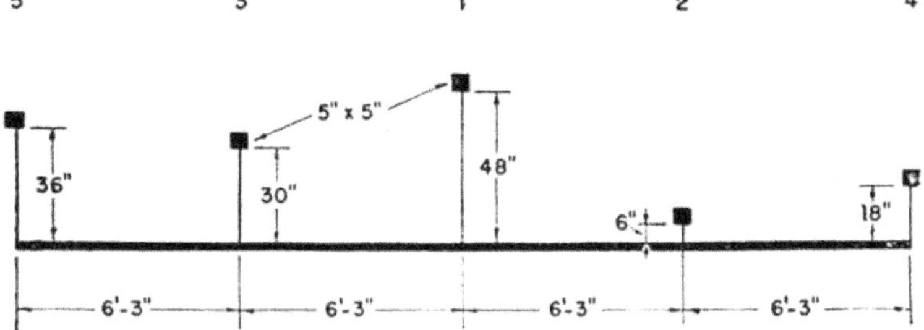

Figure 54. Manipulation target.

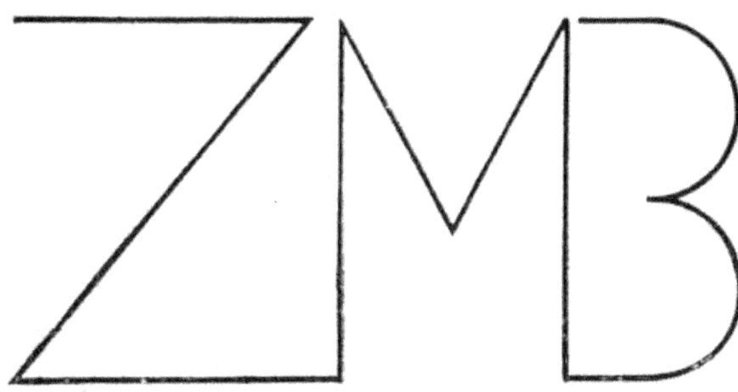

Figure 55. ZM3 character for practice in manipulation.

113. MANIPULATION. **a.** The second step in gunner training is learning the use of the elevating and traversing handwheels. Practice in their use is called "manipulation." *The gunner must know automatically what type of movement is imparted to the gun by any movement of either handwheel.*

b. The first exercise in manipulation requires the gunner to lay rapidly on a series of stationary targets. The targets should be fairly close to one another and at varying heights. This exercise is used to advantage when combined with 1,000-inch firing. Figure 54 shows a target which can be used for this type of training. The gunner lays on the left-hand target. At the command FIRE, he traverses to the center square and fires one round. He then fires one round at each of the remaining four targets in the order in which they are numbered. Time is counted from the command FIRE until the last round is fired. The course should be completed in 25 seconds. It should be fired four times, twice with manual traverse and twice with power traverse, where applicable.

c. After completing the exercises outlined in b above, the gunner should be given practice in muzzle writing with the ZM3 or "snake" board.

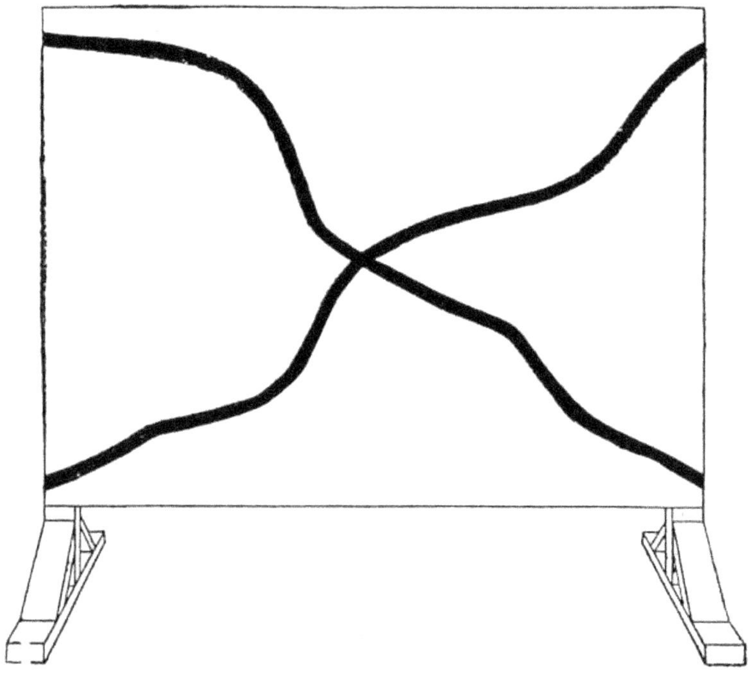

Figure 56. "Snakeboard" for manipulation practice. The standard is made from scrap lumber and target cloth. Lines are painted about 2 inches wide.

The lines (figs. 55 and 56) are painted on the side of a building, or on plywood placed on a movable standard.

(1) A block of wood holding a pencil is placed in the muzzle of the gun.

(2) A sheet of paper on a standard is then placed in front of the muzzle. (See fig. 57.)

(3) The gunner tracks the large figure with his sight. It is reproduced by the pencil on the

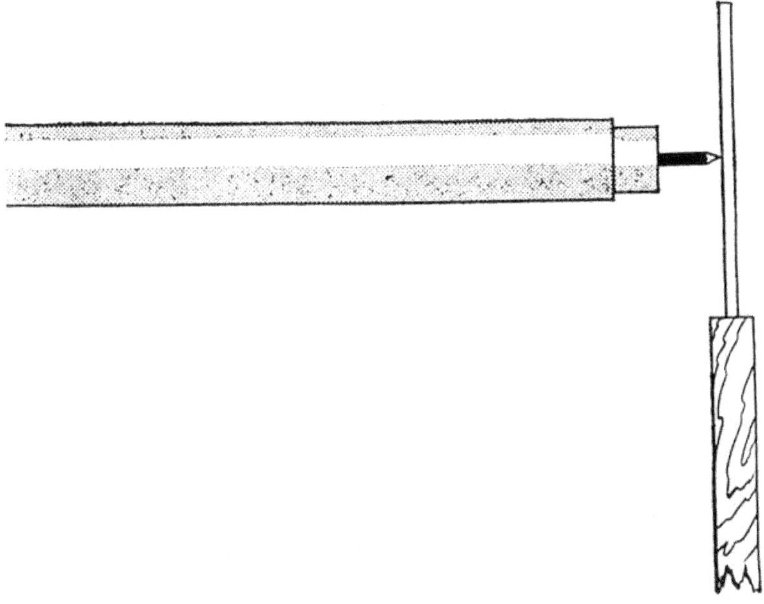

Figure 57. Standard is placed in front of muzzle of gun.

sheet of paper. Any irregularities in tracking are seen readily by studying the figure drawn on the sheet of paper.

d. After he has mastered this exercise, the gunner should proceed to exercises on moving targets. The first exercise should be easy and each successive one more difficult. The following exercises are satisfactory in the sequence listed. The gunner should be required to track a moving target—

(1) At a constant speed over a level course, maintaining a prescribed lead.

(2) At a constant speed while maintaining a prescribed range and lead.

(3) At varying speeds while maintaining a prescribed range and lead.

(4) On a hilly course which requires changes in elevation while maintaining a prescribed range and lead.

(5) While simulating adjustment of fire by executing subsequent fire orders.

e. To keep the gunner in practice, in all phases of training a short daily period of manipulation exercises should be held in conjunction with crew drill.

114. 1,000-INCH FIRING. a. Having learned manipulation, the gunner must learn to track. Tracking is merely the act of maintaining a constant sight picture on a moving target. Firing on the 1,000-inch range is the best check of the ability of a gunner to track. Essentially, 1,000-inch firing has a fourfold purpose.

(1) To teach uniform gun pointing.

(2) To teach manipulation.

(3) To start the development of the crew commander-gunner team.

(4) To teach the gunner to adjust his shots in accordance with the fire orders of the crew commander.

b. Procedure on the 1,000-inch range is as follows:

(1) The gunner is required to fire a satisfactory shot group on a stationary target. The gun is thrown off for both elevation and deflection after each round is fired. (If a 1,000-inch range is not available, the exercise described in paragraph 115 may be used.)

(2) The gunner is required to fire a satisfactory shot group on a moving target on a level run.

(3) The gunner is required to fire a satisfactory shot group on a moving target on a hilly run.

(4) The gunner is required to adjust fire as directed by the crew commander's orders.

c. The firing described in a and b above must be carefully supervised if the maximum of benefit is to be derived. Training should proceed to the firing described in b(4) above only when the gunner has clearly demonstrated his ability to perform the steps b(1) through (3) above.

(1) The goal of this firing is to teach the gunner to shoot a shot group on a moving target. The ability to shoot such a shot group is concrete evidence of ability to track a moving target.

(2) If the gunner fails to shoot a satisfactory group, the critique must bring out the reasons for his failure. The location of the initial shots with respect to the scoring silhouette will often show the source of the gunner's error.

(3) For example: If the crew commander notices that the first two shots are in rear of the scoring silhouette, he knows immediately that

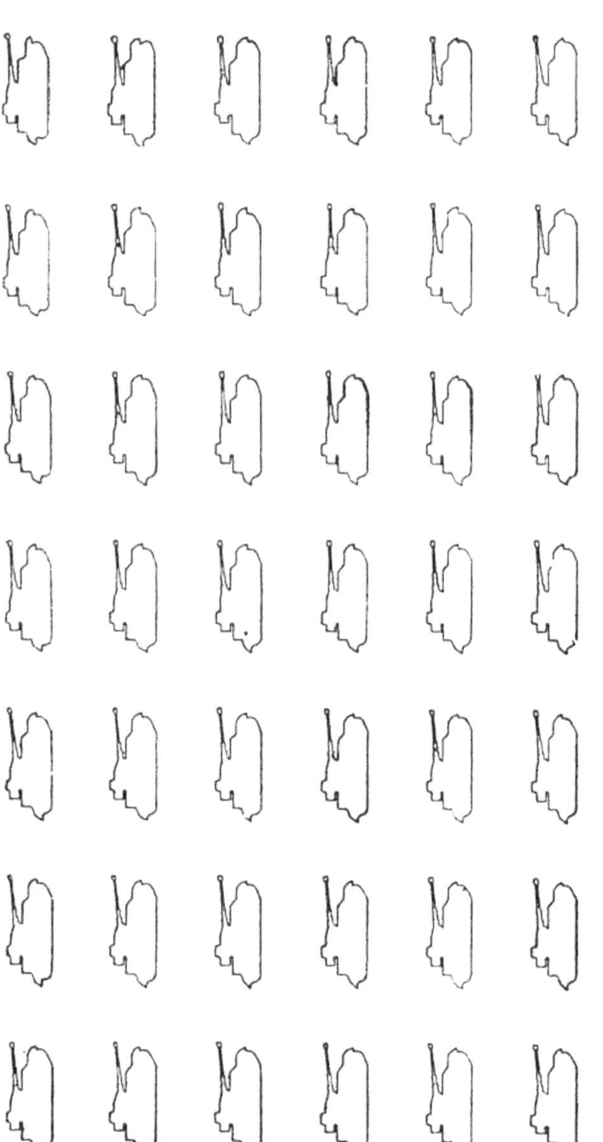

Figure 58. Calibrated 1,000-inch target.

either the sight picture was wrong or that the gunner stopped tracking at the instant of firing. A study of the gunner's manipulation technique will determine the immediate cause of the error.

d. 1,000-inch targets. (1) Standard 1,000-inch targets (FM 23–75) can be used to good advantage in the firing described in the preceding paragraphs. Even better results can be obtained by the use of targets similar to that shown in figure 58.

(2) Special 1,000-inch targets are calibrated for each type of sight. The distance between the horizontal rows of scoring spaces is equal to a change of 400 yards in range. The distance between vertical rows of scoring spaces is equal to one lead. The horizontal and vertical displacement between silhouettes can be best determined by shooting against a piece of paper. The paper should be the size of a standard target, and should be placed at a range of 1,000 inches. After the initial tests, targets can be prepared quickly by using a cardboard template to outline the scoring spaces. The scoring spaces should be drawn in lightly with a lead pencil so that they will not be visible to the gunner.

(3) Sights are adjusted as described in paragraph 50.

(4) The gunner always uses the center of the single dark silhouette as his point of aim.

115. SUBCALIBER SHOT GROUP EXERCISE. **a.** Either a subcaliber mount or a coaxial machine gun may be used. Ammunition should be all tracer, fired

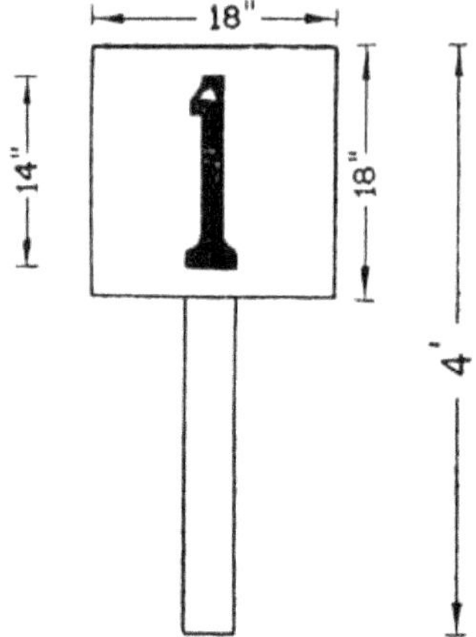

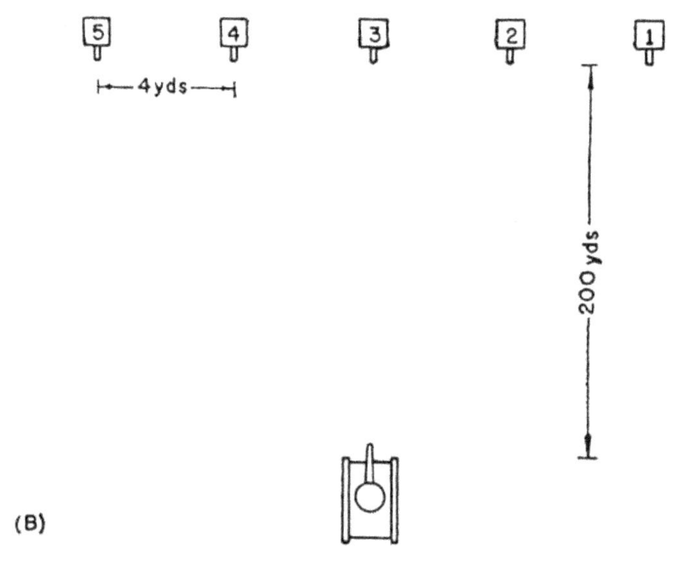

Figure 59. Target and range for subcaliber shot group firing.

single shot. This exercise is a substitute for the exercise described in paragraph 114.

b. The target is a cloth or cardboard target, 18 inches square. (See fig. 59.) It is placed 200 yards in front of the gun. Targets should be numbered to insure proper identification.

c. The sights are adjusted as for normal range firing. The gunner is informed of the range. Fifteen single shots are fired. Adjustment of each shot is by observation of the tracer. After each shot, the gunner moves the gun at least 15 mils off the target in both elevation and deflection. He relays, using the appropriate sight picture, and fires again.

d. The critique should cover the success or failure of the gunner in obtaining a satisfactory shot group.

Section III. CREW COMMANDER TRAINING

116. INTRODUCTION. The first phase in training the crew commander covers instruction in fire orders and sensings. This instruction must be accompanied by practice in range and lead estimation (chs. 2 and 3) and in target designation.

117. TARGET DESIGNATION. Crew commanders should be given practice in designating targets both when posted on the gun carriage and when posted off the gun carriage.

a. Practice in designating targets when posted on the gun carriage is conducted with the use of the TRAVERSE RIGHT (LEFT) method of designating "Direction." Where applicable, the vane sight should be used. The following subcaliber firing exercise is excellent for use with the vane sight.

(1) All sights are adjusted by boresighting. The telescopic sight aperture is covered.

(2) The instructor indicates a target to the crew commander.

(3) The crew commander gives an initial fire order. The gunner executes the fire order, stopping the traverse at the command ON, and setting an elevation corresponding to the announced range. The elevation for the machine gun is obtained from the aiming data chart or firing table. A round of tracer is fired. The horizontal deviation is measured and announced by the crew commander. His measurement is checked by the instructor. A man can be considered to have laid the gun satisfactorily if the measured deviation does not exceed 4 mils.

b. When firing high velocity guns, it is frequently necessary for the crew commander to dismount from the vehicle to observe his fire. When dismounted, the problem of target designation is often difficult. Targets may be camouflaged or otherwise made difficult for the gunner to identify. To assist the gunner in identifying the target, crew commanders must be expert at designating targets. Two excellent aids for training in this subject are described below:

(1) *Terrain plot.* Using a terrain plot, or a vantage point on the terrain, the crew commander designates a target. The gunner lays on the tar-

get using an alidade or similar device. The instructor then checks to see if the gunner is laid on the target designated by the crew commander.

(2) *Landscape target.* This training aid (fig. 60) requires the use of a 1,000-inch range, subcaliber equipment and ammunition, and two 1,000-inch landscape targets. (See FM 23–10.) The crew commander, using the landscape target immediately in front of him, designates a target to the gunner. The gunner selects a target fitting the description and fires at it. The bullet hole in the target should indicate whether or not the crew commander designated the target accurately.

Figure 60. Use of landscape targets for training in target designation.

118. TRAINING IN CONDUCT OF FIRE. The initial stages of training in conduct of fire may be taught in the form of blackboard drill. Paragraphs 119, 120, and 121 describe training aids which are excellent for instruction in conduct of fire. *If time* and equipment do not permit construction of elaborate devices, the training can be given with a twig on a bare patch of earth.

119. TERRAIN BOARD. a. The terrain board (fig. 61) is constructed as follows:

(1) A painted 6′ x 4′ board is mounted horizontally about 3 feet above the floor.

(2) An aiming circle or binoculars are placed at the same height and 400 inches from the center of the board.

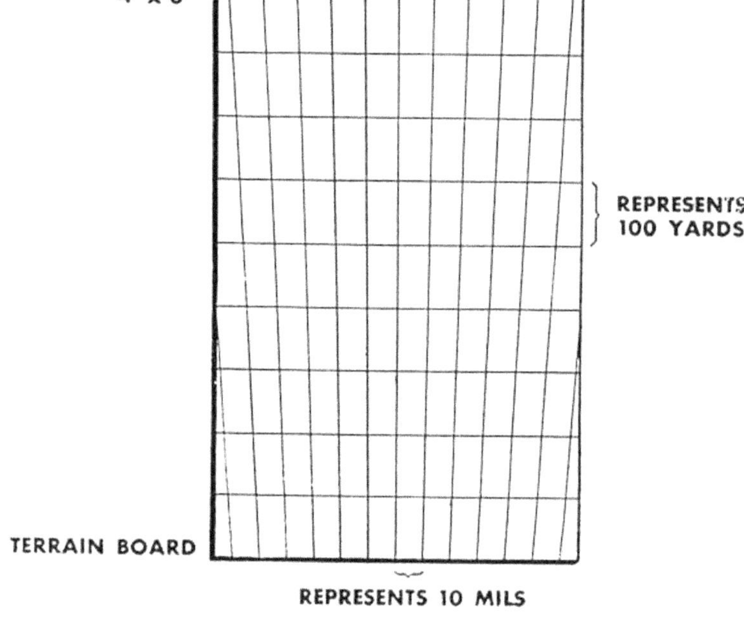

Figure 61. Terrain board showing range and deflection lines.

(3) The longitudinal center line of the board is determined and sketched in lightly.

(4) The aiming circle or binoculars are positioned so that the line of sight through the vertical center of the reticle is along the longitudinal line previously determined as in (3) above.

(5) Successive angles of 10 mils are measured right and left from the center line, and deflection lines are sketched in.

(6) Beginning at the front edge, parallel range lines are drawn across the board perpendicular to the center line. The distance between range lines is 8 inches.

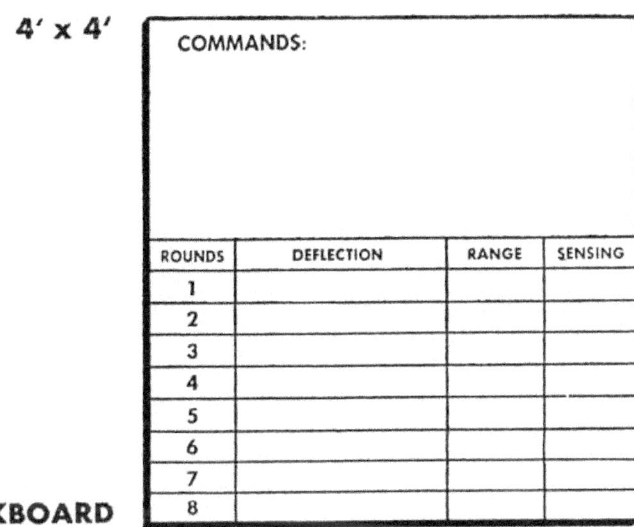

Figure 62. Blackboard to record commands and sensings.

b. Additional equipment is required as follows:

(1) Blackboard and chalk to record the initial fire order, sensings, and subsequent commands. (See fig. 62.)

(2) *HE and smoke spotter.* This is constructed by fastening two sheet metal semicircles, with a radius of ¾ inch, to the ends of a 3-foot heavy wire rod. One semicircle is painted to represent the burst of HE. The other metal semicircle is painted to represent the burst of smoke shell. (See fig. 63.)

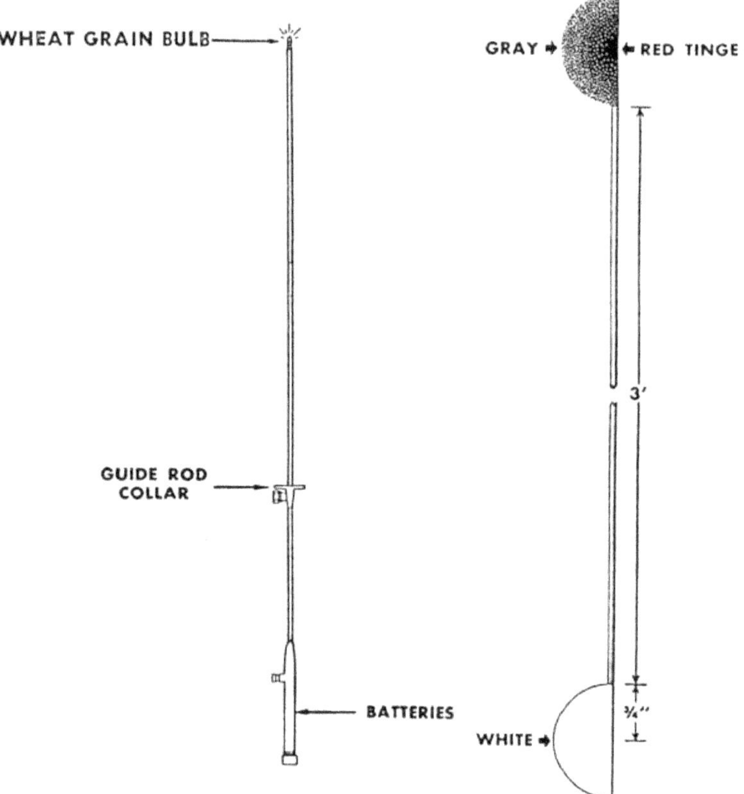

Figure 63. Accessories for terrain board.

(3) *Shot pointer.* This pointer is fitted with a small red light which is activated by flashlight cells contained in the handle. The red light is used

to represent the path of the tracer element in firing shot problems. (See fig. 63.)

(4) *Pointer guide.* The distance between the guide rods is 1 inch. Each pair of alternate guide rods represents a vertical change of 5 mils. They are brought together at the open end for ease in identification. Figures 64 and 65 give construction details of a suggested pointer guide. The base and back of the guide should be made to resemble a terrain feature.

(5) *Targets.* Targets are made in miniature. They should include enemy tanks, antitank guns, pillboxes, and fortified positions. Tanks should be constructed to scale to give an exact representation of the target at the range selected.

c. Training procedure is as follows:

(1) The individual firing the problem sits on a chair placed 400 inches from the center of the terrain board. (See fig. 66.) This position corresponds to that of the aiming circle.

(2) The instructor, acting as the terrain board operator, assigns the range lines values in multiples of 100 yards. The range values are not announced to the class. The target is announced to the student. The range to the target may be indicated by any of the following methods. They are listed in order of desirability.

(*a*) Tank targets are constructed to scale, and the student is required to determine the approximate range by application of the mil relation. (See sec. III, ch. 2.)

(*b*) The ranges represented by the near and far edge of the terrain board are indicated, and the

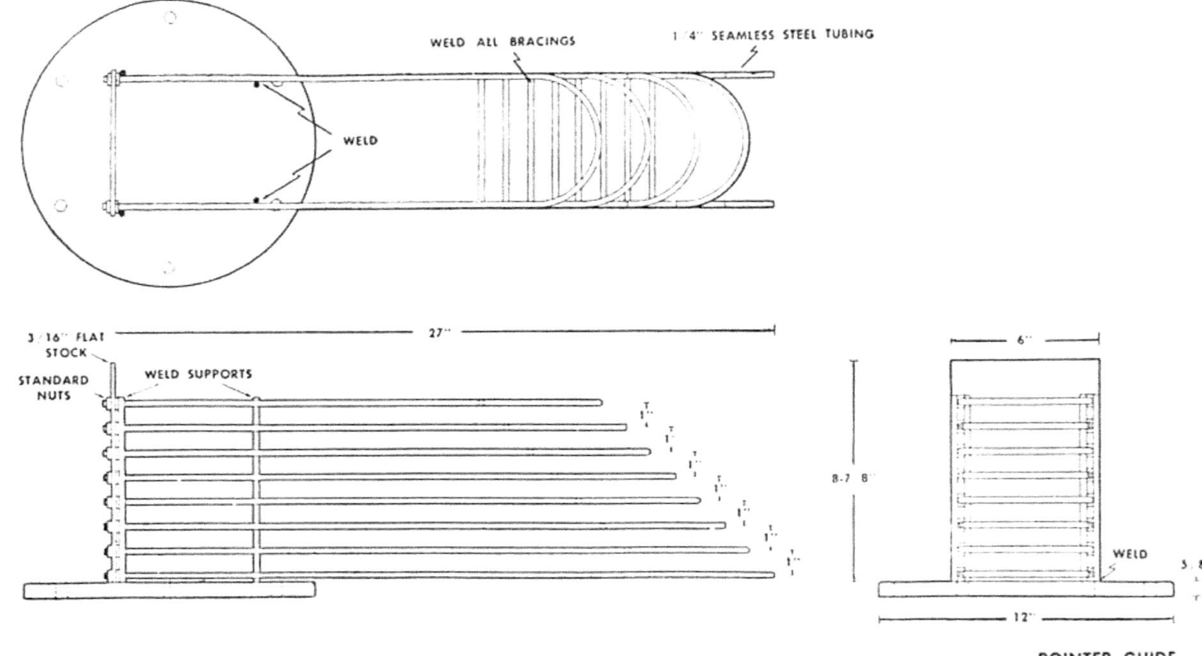

Figure 64. Details for construction of the pointer guide.

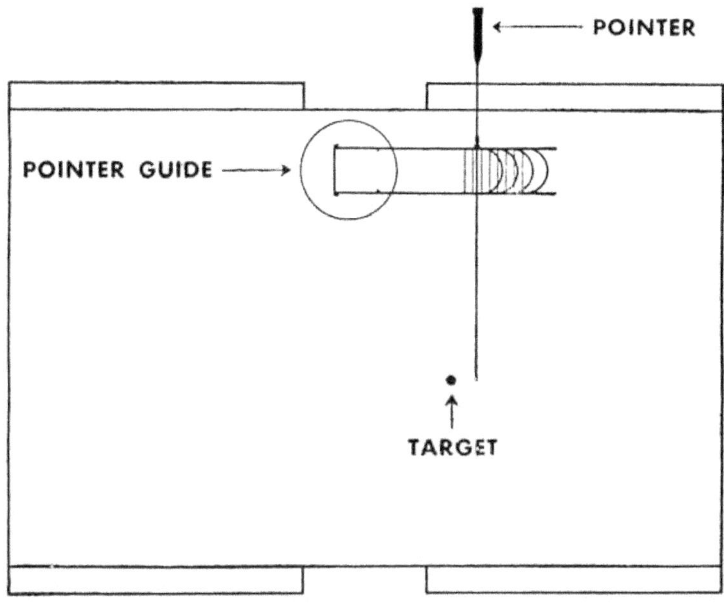

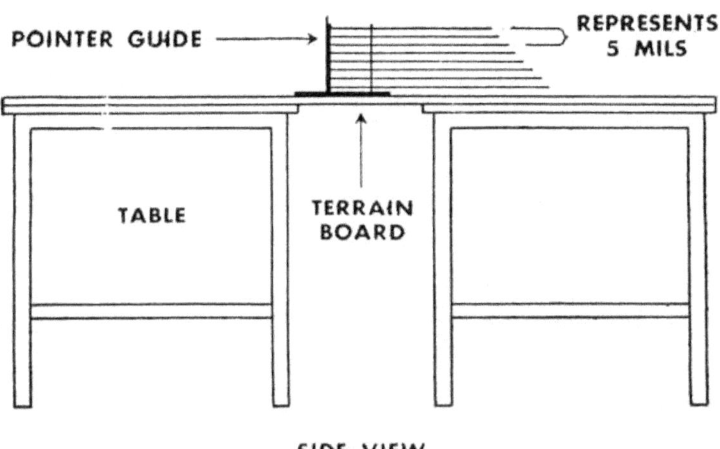

Figure 65. Assembly of pointer guide.

student is required to determine the range by interpolation.

(*c*) The gun-target range is announced to the student.

(3) The individual firing the problem gives an appropriate initial fire order. At the completion of the fire order, the instructor calls ON THE WAY and then marks the burst, strike, or path of tracer as follows:

(*a*) *HE or smoke burst.* The appropriate spotter is moved along the terrain board by means of the wire handle. It is kept flat until it has reached the point where it is desired to indicate the burst. It is then given a quick flip to the vertical, held there momentarily, then turned down slowly.

(*b*) *Path of tracer.* With the pointer guide placed on the terrain board, open end away from the student, the pointer is inserted so that it rests evenly on the appropriate pair of guide rods. With the bulb lighted, the pointer is moved so that the light passes over or near the target. The movable guide rod collar permits adjustment in deflection. To indicate a "short," the lighted tip of the pointer is dropped in front of the target. With a little practice, the instructor will be able to simulate the ricochet of shot in front of the target. The calibrated guide rod permits the instructor to give the student exactly what he calls for in subsequent orders.

(4) The initial fire order, sensings, and subsequent commands will be recorded on a blackboard as shown in figure 62.

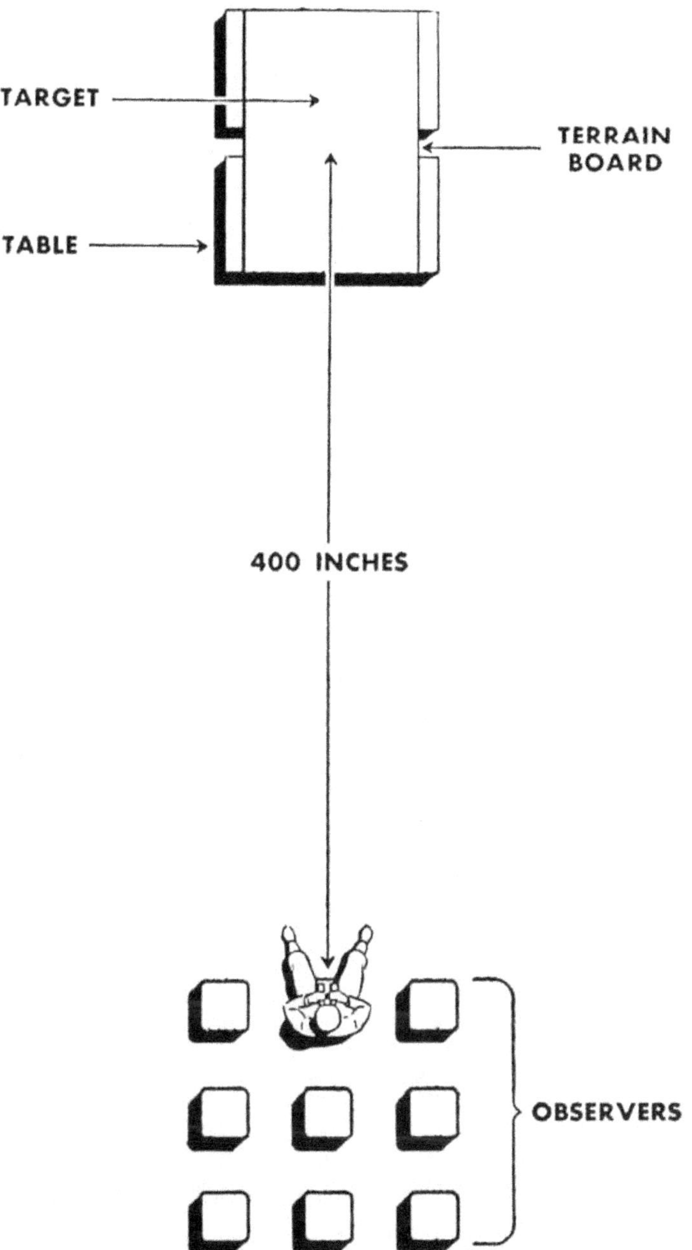

Figure 66. Setup for instruction with terrain board.

(5) The conduct of the critique of a problem is discussed in paragraph 5.

(6) Some practical hints for instructors in conducting terrain board problems are:

(*a*) The first round should seldom hit the target. In actual combat firing, there will be errors caused by gunners, cant, variations in jump, improper sight adjustment, and other factors.

(*b*) The shift indicated by the subsequent fire orders should not always be given.

(*c*) The target is always designated as an enemy vehicle or installation, using the correct nomenclature.

(*d*) The problems should be varied to illustrate the proper selection of ammunition and methods of fire adjustment.

120. TERRAIN FAN. This training aid (fig. 67) is not as elaborate as is the terrain board. Further, the absence of the guide rails requires the instructor to demonstrate a thorough knowledge of trajectory. The simplicity of the terrain fan, however, permits it to be constructed easily on the ground or painted on a piece of plywood and moved from place to place.

a. The terrain fan is designed for practice in conduct of fire with shot. With a slight change in technique, it can also be used to train for conduct of fire with HE or smoke.

b. The fan is used for training in conduct of fire when the distance that the round strikes short of the target can be estimated. (See par. 102a.) Exercises should also be conducted on a fan from

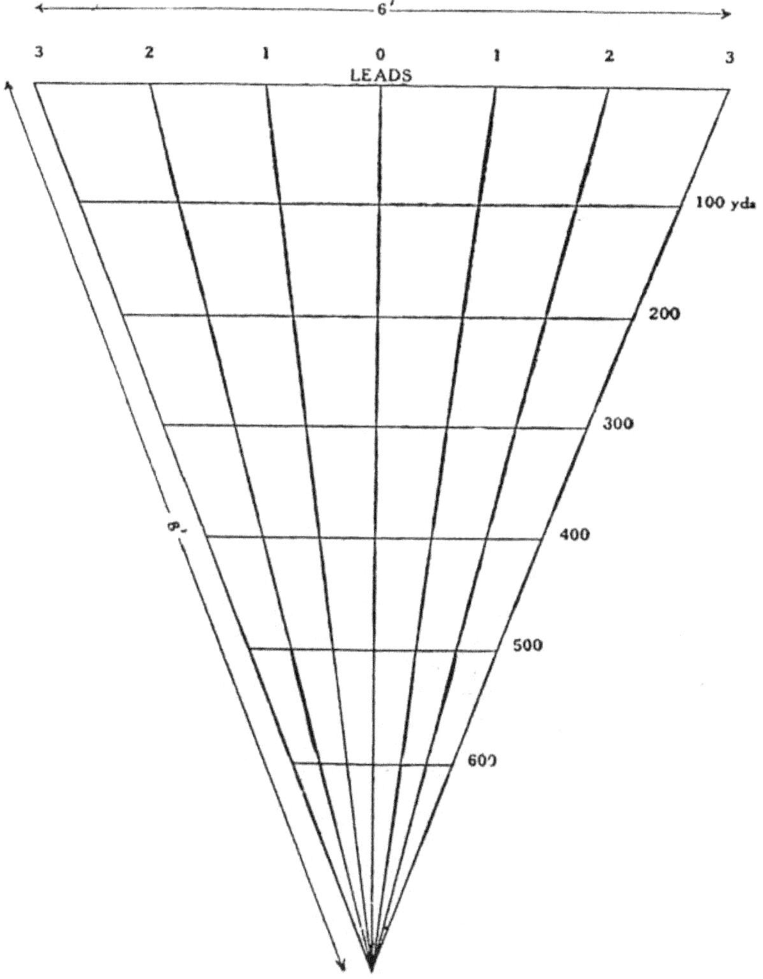

Figure 67. The terrain fan.

which the range lines have been removed to give practice in the technique of adjustment when the error cannot be estimated.

c. To use the fan properly, a model tank, a blackboard, a large sight reticle, a mounted tank silhouette, and a pointer equipped with a red tip are needed.

(1) The platoon leader stations one man in the crew commander's position, another man at a blackboard which should be placed in front of the students, and a gunner at the sight reticle with his back to the remainder of the class.

(2) The platoon leader then places the tank model on the fan at the zero range, zero lead mark, and announces that the tank is 1,200 to 1,600 or 800 to 1,200 yards distant, or whatever range he decides to use. He then announces whether or not the tank is moving. If moving, he announces its "apparent speed."

(3) He then calls on the crew commander for a fire order. As the crew commander gives the fire order, the man at the blackboard records it in the spaces provided. At the proper moment, the gunner announces ON THE WAY.

(4) The platoon leader stops the problem and checks with the class to see if the fire order was given correctly and if the gunner set off the correct sight picture.

(5) He indicates, using his pointer, where the round landed or passed the tank. He then requires the crew commander to give a subsequent fire order and the gunner to set off the new range and deflection. To indicate an "over," the platoon leader moves the red tip of his pointer over the target to simulate the tracer in the round.

(6) With a little ingenuity, the platoon leader can reproduce almost any situation his men may face on the battlefront.

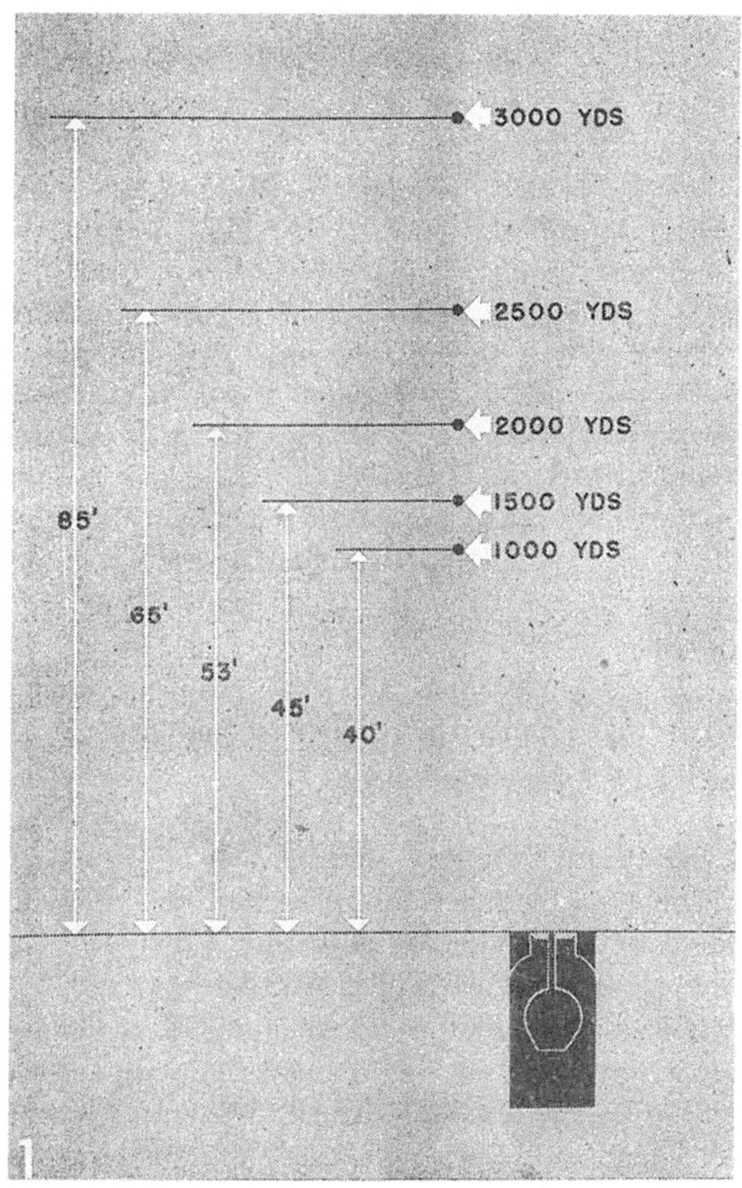

Figure 68. Plan view of range for fire adjustment trainer.

d. A suggested basic use of the trainer follows:

(1) The crew commander gives a fire order. The gunner announces ON THE WAY. If the fire order and the gunner's sight picture are correct, a hit is given.

(2) The crew commander should then shout TARGET, and the gunner ON THE WAY, for the second round with the same sight picture. If this is done, the platoon leader should announce that the tank is hit on the track and is stopped.

(3) The crew commander should then command ZERO LEADS, and add the range change required to move the strike to the center of the target. The gunner sets off the new sight picture. If the subsequent fire order and sight picture are correct, the platoon leader indicates another hit.

(4) After conducting a few problems in the preceding manner, instruction should proceed as outlined in paragraph 121c.

121. FIRE ADJUSTMENT TRAINER. a. Training of the crew commander in conduct of fire may be aided by use of the fire adjustment trainer. Its construction and use is discussed below.

b. For the range (figs. 68 and 69), a flat or fairly flat piece of terrain covered with either short grass or sand is selected. Miniature roads, railroads, buildings, and other features may be added but are not necessary. Ranges are indicated by marked stakes.

Figure 69. Panoramic view of range for fire adjustment trainer.

Figure 70. Mount for submachine gun on 75-mm M3.

c. Subcaliber mounts are adjusted as follows:

(1) The mounts shown in figures 70 and 71 may be used for the 75-mm and 76-mm guns mounted on the M4 tank. Modifications thereof may be designed for other gun carriages.

(2) The solenoid, which fires the coaxial machine gun, is dismounted, installed on the submachine-gun bracket (fig. 72), and connected to

Figure 71. Mount for submachine gun on the 76-mm gun.

its original position by an extension wire. Care must be taken with the wiring to prevent a short circuit in the electrical system.

(3) *To make adjustment for 75-mm gun.* (*a*) A reading of 22 mils is set on the M9 quadrant and the bubble leveled with the handwheel. By using the battle sight on the submachine gun without moving the tank gun, it is adjusted to hit the base of the stake at the 1,000-yard marker. The clamps are tightened.

Figure 72. Installation of solenoid on submachine gun.

(*b*) In setting 22 mils on the M9 quadrant two factors have been considered. 10 mils of the 22 mils is the firing table elevation for the adjusting-in range of 1,000 yards. The remaining 12 mils introduces a calculated error which, when the firing table elevation for the target concerned is applied, will make the initial round fall short the range-change equivalent of 12 mils. Based on a "c" of 4 mils for ranges up to 2,000 yards, the first round will be approximately 300 yards short, thereby requiring bracketing.

(*c*) The trigger arm between the solenoid and the submachine gun trigger is adjusted for proper operation, and the submachine gun is set on "single."

(4) Typical targets are small chips, caliber .45 cartridge cases, and match boxes. Some targets are placed in inconspicuous positions where they can be observed only with field glasses.

d. The observer acts as crew commander while another member of the crew performs the duties of the gunner. The loader handles the submachine

Figure 73. Position of personnel when using fire adjustment trainer.

gun and initially lays the gun for deflection. On the first round, a deflection error of from 10 to 20 mils is made in the laying of the gun to afford practice in measuring deviation. The crew commander and instructor, with the other observers, assume prone positions under the gun carriage.

(See fig. 73.) All are equipped with field glasses. The procedure is as follows:

(1) The instructor designates a target.

(2) The crew commander issues a fire order.

(3) The gunner executes the order, using a special firing table ((5) below) and the M9 quadrant.

(4) The crew commander observes fire and gives the necessary commands for range and deflection changes. Range changes are converted to mils and applied to the elevating handwheel or elevation quadrant by the gunner. Deflection changes are made with the azimuth indicator.

(5) A firing table is prepared using an elevation of 10 mils for 1,000 yards. From 1,000 to 2,000 yards, a "c" of 4 mils is used, and from 2,000 to 3,000 yards, a "c" of 5 mils is used.

Section IV. TEAM TRAINING

122. GENERAL. **a.** Although the final phase of 1,000-inch firing and the use of the "fire adjustment trainer" require a certain amount of coordination between the gunner and crew commander, this coordination reaches its height in field firing.

b. The purpose of field firing is to perfect the application of gunnery technique and to develop a smooth-working team capable of conducting fire under combat conditions. Emphasis is placed on sensing and adjustment.

c. Firing lines should be arranged to permit the maximum of control and of ease in the supervision of instruction. Initial firing exercises are executed without cover, concealment, or camouflage.

As the crews progress in their training, firing from covered, dug-in, and camouflaged positions is included.

123. FIELD FIRING. Field firing should start with stationary target firing and progress to firing at surprise and moving targets. It may be divided into subcaliber firing and service firing.

a. Subcaliber field firing. This initial stage of field firing provides instruction in laying, tracking, and the application of the technique of gunnery without the disturbance of muzzle blast and the shock of recoil.

(1) Firing on a moving tank with caliber .30 or .50 ammunition is the best subcaliber training a gunner can receive.

(2) In the absence of tanks, towed panel targets may be used for both caliber .30 and .50 subcaliber firing.

b. Service firing. (1) The importance of service firing cannot be overemphasized. This firing is the culmination of all previous gunnery training. Only those men who have passed the gunner's proficiency test and have completed satisfactorily both 1,000-inch firing and subcaliber field firing should be allowed to fire.

(2) Service firing should be conducted at both stationary and moving targets. To obtain the maximum gunnery training from ammunition expended, gun crews must work as a unit, and every problem must be critiqued. The object of service firing is to perfect the gunners and crew comman-

ders. It should present every possible battlefield target that range facilities will allow and that imagination and ingenuity can devise. Realism should be stressed and uniformity avoided.

(3) The firing should test the ability of the crew commander and gunner to select the proper ammunition and fire upon *any target* that may face them on the battlefield, not simply their ability to hit a target whose direction and location they know. Firing should progress from close to more distant ranges, and from plainly visible to partially visible targets. It should include shifting from moving targets to surprise (disappearing or "flipper") targets.

124. STABILIZER FIRING. a. This exercise should be performed by crews training on gun carriages equipped with gyrostabilizers and coaxial machine guns. It provides training in manipulation, use of the sight, and observation of fire while moving. (See sec. V, ch. 14.)

b. The gyrostabilizer and power traverse are used. Sights are adjusted as for normal range firing. Single shot tracer is fired with the coaxial machine gun. Each man should "dry run" the course many times.

c. Terrain is selected over which the driver can maintain an average speed of 12 mph. A course about 1,500 yards long is laid out. Targets are placed at the following ranges from the starting point: 300 yards, 600 yards, 800 yards, and 1,300 yards. Targets are 3′ x 5′ panels painted olive

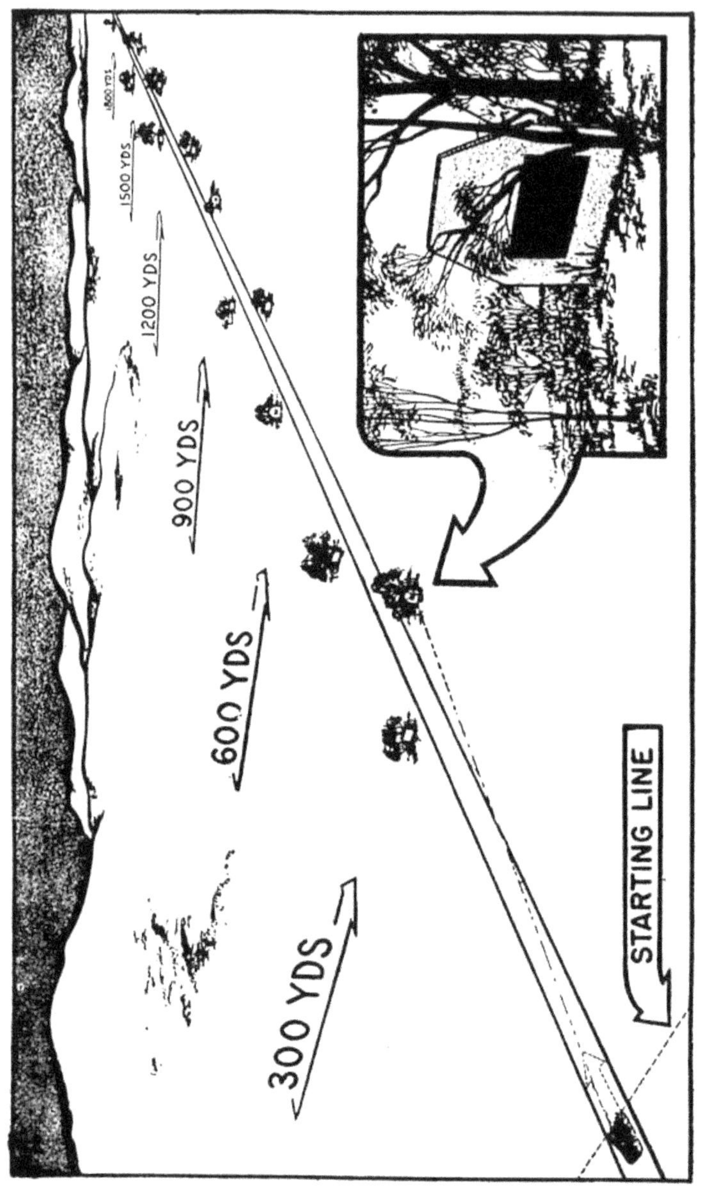

Figure 74. Target and range for stabilizer firing.

drab, or a similar color, and tank silhouettes as shown in figure 74. Some targets are placed on each side of the runway. Gunners are not permitted to fire when they are less than 50 yards from the targets.

CHAPTER 14

ADDITIONAL METHODS OF FIRE

Section I. NIGHT FIRING

125. GENERAL. a. Night firing presents three situations:

(1) Firing when the target is illuminated by bright moonlight or by artificial means.

(2) Firing at targets which disclose themselves at unlocated positions.

(3) Firing during darkness at targets previously located during daylight.

b. When possible, fire should be observed and adjusted.

c. In direct fire at night, emphasis must be placed on speed. Alternate positions must be selected and prepared. Positions must be occupied rapidly, fire delivered, and guns displaced to alternate or cover positions. Careful preparation is essential.

126. ILLUMINATED FIRE CONTROL INSTRUMENTS. In all situations, illumination is necessary for sights, azimuth indicators, and aiming circles. Night lighting devices with normal brightness give too much light in the sight to permit the observation of machine-gun flashes. This can be overcome by using weak batteries or by partially covering the bulb.

127. ILLUMINATED TARGETS. Targets may be visible at night due to natural light of a bright moon or through use of artificial illumination, such as illuminating shell, flares, fires, and searchlights. With the target illuminated, pillboxes, self-propelled weapons, and machine guns may be identified on the terrain. Illuminating shell should be fired only to locate a target accurately after an approximate location has been obtained. To prevent their exposure it should be used only after consultation with supported troops. In some cases, flares or mortar illuminating shell may be used by the infantry to illuminate the target for the supporting troops. Close coordination is necessary. Haystacks, wooden houses, and other inflammable material may be ignited by the use of tracer and white phosphorous ammunition. Care must be taken to avoid setting fires when the light therefrom will disclose the location of friendly troops. Searchlights, using either direct or indirect lighting, are the most satisfactory source of battlefield illumination. Tactical employment of the searchlights should be the responsibility of the commander of the unit to which the searchlights are attached, normally the division commander. *When the target is illuminated, normal methods of laying and adjusting fire may be employed.* When firing at illuminated targets, data should be recorded.

128. TARGETS OF OPPORTUNITY. Targets of opportunity may be revealed by muzzle flash of enemy weapons. *Daylight methods of laying are used.*

The range may be determined by any of the methods listed in paragraph 8. The most accurate means permitted by time and tactical considerations should be used. The procedure is as follows:

a. Lay on the flash, using the coaxial telescope and the proper range line. The quadrant elevation is recorded.

b. The reading on the azimuth indicator is recorded or an aiming point for the panoramic sight is selected for use if the enemy weapon stops firing.

c. Fire is adjusted by the methods described in paragraph 131.

129. DESIGNATION OF TARGETS BY INFANTRY. a. The infantry may designate targets by using two flank machine guns. (See FM 17–36.) These guns are fired so that their tracer streams intersect directly above the target or on the target itself.

b. Daylight methods of laying are used. The point of aim is determined as follows:

(1) If the tracer illuminates the target, the gunner lays directly on the target.

(2) If the target is not illuminated, the gunner lays on the intersection of the tracer streams.

c. Fire may be adjusted by daylight methods, or by the methods described in paragraph 131.

130. TARGETS LOCATED DURING DAYLIGHT. When targets to be engaged at night are identified during daylight, the following procedure is used:

a. When the approximate position for the gun

has been selected, an aiming circle is set up on the gun-target line. A stake is placed directly under the aiming circle to indicate the gun position. A second stake is put in 100 yards from the piece stake. A third stake is placed midway between the two stakes and exactly in line. These stakes may be placed in any direction from the piece stake, provided the correct angle to the target is determined and recorded. Since it is necessary to put night lighting devices on the stakes, it is important that they be placed so that the enemy will not be able to observe them.

b. From the gun position, the angle of site to the target to be fired upon is obtained. When the gun can be moved into position during daylight, the angle of site may be obtained as outlined in paragraph 25b. When this cannot be done, the angle of site may be determined with an aiming circle.

(1) If an aiming circle is used, it must have been calibrated previously with the quadrant which will be used in the actual firing.

(2) If the angle of site is not measured from the same altitude as that of the gun trunnions, a correction must be computed for the difference, using the mil formula.

c. The range is converted to mils of elevation from the firing table and the angle of site is applied. This is the quadrant elevation to be fired. With area targets, fire should be conducted at this elevation and at elevations ½ "c" above and ½ "c" below. When firing at pillboxes, it is necessary to fire all rounds at the same elevation to obtain the necessary cumulative effect for penetration. (See par. 144.)

d. The gun is brought into position over the piece stake. To assist in driving into position, a white tape about 40 yards long may be stretched along the gun-target line. The gunner then lays the tube on the two aiming stakes. If necessary, the carriage is moved at the direction of the gunner until the tube or sight is lined in with both stakes.

e. The gun is traversed the angle previously determined, a above. The elevation corresponding to the range, plus or minus the necessary site correction, is set on the gun, using the quadrant. Deflection and range corrections can be made by using the aiming circle to sense the rounds. The procedure is described in paragraph 131.

f. If it is not possible or practicable to aline the tube with the stakes, a displacement correction must be applied. This is accomplished as described in paragraph 178.

(1) After correcting for lateral displacement, the gun is parallel to the former gun-target line.

(2) To move the gun onto the target, it is now necessary to measure the distance the sight is displaced from the stake. The range to the target is used and the mil formula is applied. For example: If the gun carriage were 4 yards off the stake and if the range were 1,000 yards, it would be necessary to traverse the gun 4 mils to hit the target.

131. ADJUSTING FIRE. It is not always possible to prepare positions as set forth in paragraph 130. A gun may be moved in after dark and its fire adjusted by using the one aiming circle method or the two aiming circle method.

a. In the single aiming circle adjustment, the aiming circle may be set up during daylight and its cross hairs centered on the target. During darkness, the cross hairs may be placed on the target by artificial illumination or by observing the flash. The gun is brought into position from 20–30 yards to the flank of the aiming circle. Shots appearing right or left of the vertical cross hair are right or left for deflection. The deviation may be measured directly on the instrument reticle. Shots appearing above the horizontal cross hair are over for range. Those appearing below the horizontal cross hair are short for range.

b. The accuracy of this method is limited if the ground at the target area is level or sloping gently away from the observer, as shots may strike as much as 100 yards over or short of the target although they will appear to be falling on the horizontal line of the reticle. Therefore, this system is applicable only when the target is on a forward slope.

c. As previously stated, deflection corrections are made by reading the deviation directly from the aiming circle reticle. Range is bracketed until target hits are obtained. In case of a point target upon which ricochet fire is to be used, adjustment is made with fuze set at superquick and then the range is reduced an appropriate amount before firing with fuze delay. For ranges up to 1,000 yards, the amount can be determined by dividing the muzzle velocity in feet by 4 and then multiplying by the time of delay in seconds. For example: The adjusted range for shell HE, set at superquick for the 75-mm tank gun, M3, has been determined as 900 yards. The muzzle velocity is 2,030 feet per second and fuze delay .05 seconds. Then $\frac{2030 \times .05}{4} = 25$ yards, the amount to subtract. The range will be 900 — 25 or 875 yards.

d. In an adjustment with two aiming circles, the instruments are laid on the target as prescribed for the single aiming circle method. The observer with the aiming circle located close to the gun (axial observer) adjusts the fire until the bursts are brought to the vertical hair of his aiming circle. He does not attempt to sense or adjust for range. When the deflection is correct, the flank observer continues the adjustment, sensing only for range. After an initial arbitrary range change, he can determine the relationship between the deviation measured by the instrument and the corresponding change in elevation. When the burst appears on the vertical cross hairs of both observers' instruments, range and deflection are correct and fire for effect is begun. Employment of two aiming circles eliminates the limitation imposed by using only one instrument. (See b above.) Fire can be adjusted accurately for range and deflection.

132. AMMUNITION. **a.** In night firing, flashless powder should be used when available.

b. If an adjustment is fired using an aiming circle, fuze quick should be used during the adjustment since a ricochet short may appear above the horizontal line and be sensed "over."

Section II. DIRECT LAYING TWO OR MORE GUNS

133. GENERAL. **a.** The platoon or section leader may desire to concentrate the fire of his guns on a point target or small area target, using HE. To accomplish this, he adjusts the fire of his own gun on the target and determines the data necessary for the other guns to open fire. This method conserves ammunition and is used where speed is not essential.

b. The above method is too deliberate in situations where speed is essential. For example: A tank section or platoon is advancing with hatches closed without supporting infantry, but with a section or platoon in close support. It is attacked by tank-destroying personnel teams. In such case, fire control should be decentralized to each tank commander and fire opened with machine guns. An artillery fire mission of time fire should also be requested. The fire order should be as brief as possible. For example: COVER FIRST SECTION WITH MACHINE GUN FIRE.

134. FIRE ORDERS. In an initial fire order, the platoon or section leader designates the guns to fire, the ammunition, and the target in the following manner:

NUMBERS THREE AND FOUR (PLATOON).
WATCH MY BURST.
REFERENCE POINT, RIGHT ONE THREE ZERO.
ANTITANK.
TWO THOUSAND.

(The ammunition element is omitted from the order unless the platoon leader desires the other tanks to load at this time.) He then issues the usual fire order to his own gunner and proceeds to adjust by normal direct fire methods. He announces ON THE WAY as each round is fired to alert the other crew commanders.

135. ADJUSTMENT. **a.** The commanders of the other guns order their gunners to watch the burst and have them traverse as necessary to put the burst in the field of view of the coaxial telescope. To the range announced, each crew commander adds or subtracts any estimated difference in the range between his gun and the platoon or section leader's gun. For example:

GUNNER.
TRAVERSE RIGHT.
STEADY . . . ON.
WATCH HIS BURST.
ANTITANK.
ONE NINE FIVE ZERO.

b. The gunners lay the appropriate range mark on the base of the burst effect of HE or smoke.

c. While adjusting his fire on the target, the platoon or section leader transmits ON THE WAY as each round is fired. On receiving this information, all gunners lay on the burst effect, using the initial range without further command.

136. FIRE FOR EFFECT. **a.** When the platoon or section leader obtains a target hit or establishes a 100-yard (one "c") range bracket, he commands GUNNER, RANGE. His gunner notes the range mark in the reticle at which that burst effect appeared and announces it.

b. The platoon or section leader transmits his adjusted range to the other guns and orders them to fire. The adjusted range is determined as follows:

(1) If the last round was a target, the range at which it was fired is the adjusted range.

(2) If the last round established a 100-yard range bracket, the adjusted range is derived by adding or subtracting 50 yards from the range announced by the gunner.

c. His command to the other guns is for example:

> NUMBERS THREE AND FOUR (PLATOON).
> MY RANGE TWO TWO FIVE ZERO.
> FIRE.

d. The platoon or section leader now fires one round at the adjusted range, transmitting ON THE WAY. He continues to fire on the target, without transmitting ON THE WAY, until the mission is accomplished.

e. On receiving the preceding commands, the other crew commanders order their gunners to fire. To the range announced, each crew commander adds or subtracts any difference in range between his gun and the platoon leader's gun. For example:

HE, DELAY.
TWO THREE FIVE ZERO.
FIRE.

f. The gunners lay the proper range line on the effect of the burst fired by the adjusting gun, and fire. For accurate results, it is necessary that the effect of the bursts be clearly visible.

g. If necessary, the remaining guns adjust their fire on the target. If they are unable to identify the target, they lay on the bursts of the platoon or section leader. The initial range change should not exceed 100 yards.

h. Unless the platoon or section leader designates the number of rounds to be fired, fire is continued until CEASE FIRING is given by the platoon or section leader.

137. METHOD OF FIRE. When there may be difficulty in identifying bursts, the platoon or section leader may transmit a method of fire. For example:

MY RANGE TWO THREE HUNDRED, PLATOON RIGHT, FIRE. Salvo fire is continued until the platoon or section leader commands CEASE FIRING, or commands volley fire.

He may command the number of volleys, for example:

PLATOON, THREE ROUNDS, FIRE.

138. SHEAF. A converged sheaf may be employed if the target is less than 50 yards in width. For targets from 50 to 100 yards in width, a parallel sheaf is employed. The command of the platoon leader should be an individual shift for each gun to place its burst approximately on its portion of the target. Computation of such shifts is covered in paragraph 183. The deflection shift is not executed by the gunner until after laying for range on the round fired by the platoon leader's gun at the adjusted range.

139. SURPRISE ATTACK OF TARGETS. In a surprise attack of targets, it is necessary to assign each tank a portion of the target and have all guns open fire at the same time. Such a target would be a column of personnel or vehicles on a road. For example, the initial fire command would be—

PLATOON.
CALIBER THIRTY.
MY DIRECT FRONT.
TRUCKS.
SIX HUNDRED.
NUMBER TWO TAKE FIRST TWO TRUCKS.
NUMBER THREE SECOND TWO TRUCKS.
NUMBER FOUR THIRD TWO TRUCKS.
NUMBER FIVE FOURTH TWO TRUCKS.

AT MY COMMAND.
REPORT WHEN READY.

Each tank reports in sequence, for example: NUMBER TWO READY. When all are ready, the platoon leader commands FIRE. For surprise attacks on other targets, see FM 17–30.

Section III. USE OF RANGE CARDS TO MASS FIRES

140. GENERAL. When a platoon has sufficient time in the organization of its direct fire positions, plans should be made to coordinate fires in the event of night attack or an attack through smoke. Fire is coordinated by the platoon leader in the following manner:

a. From the duplicate range card (par. 17) of each direct-fire position and from personal reconnaisance of the area, he selects probable tank approaches. Key terrain features in these approaches are given a common designation. For example: A, B, C, D. (See fig. 75.)

b. The platoon leader now sends a command similar to this:

PREPARE DATA FOR PREARRANGED FIRES ON THESE POINTS:

A—Bridge.
B—Road junction.
C—Bend in road 1,200 yards west of bridge.
D—Ford 800 yards west of bridge.
E—Left edge of heavy woods east of bridge.
F—Saddle in ridge south of position.
G—Old well.

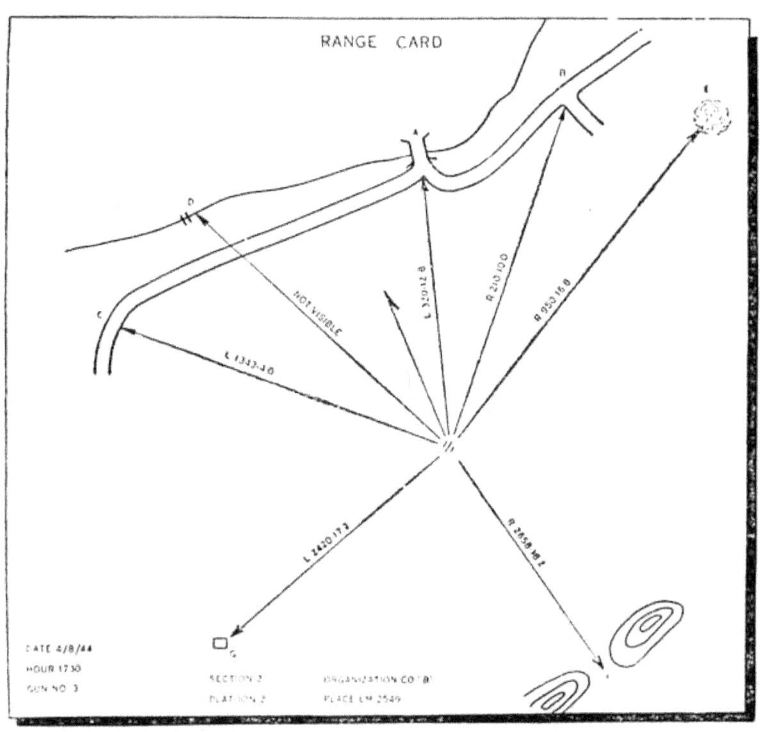

Figure 75. Range card prepared for massing fire.

141. PREPARATION. Each crew commander prepares the data for all points from his position. This is done in the same manner as in the preparation of the direct-fire range card described in paragraph 15. In addition to obtaining the data on the direct-fire range card, the crew commander—

a. Converts ranges into elevations by referring to firing tables.

b. Determines the angle of site by laying on each target with the zero range line and measuring the angle of site with the gunner's quadrant.

c. Obtains the quadrant elevation by applying a plus or minus angle of site to the firing table elevation for each target range.

d. Uses a tabular record to assist in preparation of this firing data. (See fig. 76.)

e. Shows the firing data on his range card. (See fig. 75.)

142. PROCEDURE. If it becomes necessary to employ massed fire, the platoon leader, from a point of vantage, controls the massing of fire of the guns by a command such as TARGET F, FIVE VOLLEYS, AT MY COMMAND. Each crew commander who can place fire on this target sets off the prearranged data and reports NUMBER ______READY. When all guns have reported,

TABULATION OF DATA					
CO "B" 2D PL GUN NO.3		PREARRANGED FIRES			REFERENCE POINT AIMING POST
TARGET	ANGLE FROM REFERENCE PT	RANGE	ELEVATION FROM FIRING TABLE	ANGLE OF SITE m	QUADRANT ELEVATION
A	L 320	1550	98	+30	128
B	R 210	1800	120	−20	100
C	L 1343	1400	86	−46	40
D		NOT VISIBLE FROM GUN POSITION			
E	R 950	1900	128	+40	168
F	R 2658	1700	110	+72	182
G	L 2420	2150	150	+22	172

Figure 76. Data sheet for massing fires.

the platoon leader commands: FIRE. If the platoon leader desires each gun to fire when ready, his initial command is TARGET F, FIVE VOLLEYS, FIRE. HE ammunition is used in this type of firing.

Section IV. PRECISION FIRING

143. GENERAL. High velocity guns are extremely effective for engaging point targets such as concrete pillboxes or caves. The relatively small amount of vertical dispersion facilitates the reduction or neutralization of points of resistance of this type. Penetrating the walls of a concrete pillbox is similar to digging a tunnel. Each shovelful must be taken from the same hole. The procedure used in destroying or rendering ineffective a cave which is being used as a strong point depends upon sealing the opening by causing the roof to cave in. In both instances, the result is achieved through the cumulative effect of several rounds fired into the same spot. When closing the entrance to a cave, the rounds are fired at points above the opening. A slight deflection shift may be necessary to obtain the desired effect. The utmost accuracy in determining firing data and in laying is required.

144. METHODS OF ATTACK. **a.** Targets such as those mentioned in paragraph 143 may be attacked by direct or by indirect fire methods. To utilize the full potentialities of the gun, ranges should be as short as the terrain and the tactical situation will permit. Wherever practicable, positions should be selected so that the line of fire is perpendicular to the face of the target. If the cave entrance is so situated that rounds cannot be fired at or directly above it, a channel to the entrance may be cut by firing. If excessive ricochets occur, a surface which will provide sufficient retardation to hold subsequent rounds may be produced by HE shell with superquick fuze or by concrete piercing shell with nondelay fuze.

b. No matter which method of attack is employed, fire for effect can best be conducted under the surveillance of a forward observer. A forward observer can get closer to the target, and can—

(1) Better identify the target.

(2) Adjust fire more accurately.

(3) Determine when destruction or neutralization has been accomplished.

c. When the target is visible through the coaxial telescope, direct fire methods may be employed. Laying and adjusting are expedited when using the reticle. Unless the target is temporarily obscured by smoke or dust, this method is generally faster than is the indirect fire method. The procedure is described in paragraph 145.

d. Attack of caves and pillboxes by indirect fire methods (par. 146) has the following advantages:

(1) Targets can be attacked which cannot be seen from any potential direct fire positions.

(2) Fire may be delivered from a defiladed position.

(3) Fire can be continued without interruption after the target becomes obscured.

(4) The main disadvantage is that more time is required for preparation of data and initial laying.

e. When using small caliber guns such as the 76-mm, two guns may be assigned to each target.

145. ATTACK BY DIRECT LAYING METHODS. **a.** Lay on the target with the direct fire sight, using the most accurate range available. At longer ranges, it may be desirable to correct the deflection for drift and wind.

b. Adjustment should be continued until the center of impact is on the target.

c. (1) If the target is obscured during the latter phases of adjustment, the gunner should use the gunner's quadrant to lay for elevation, to the nearest tenth of a mil when practicable.

(2) To repeat the correct deflection setting when the target is obscured, the following two methods can be used:

(*a*) Select an auxiliary aiming point within the field of view of the coaxial or panoramic sight.

(*b*) Determine the azimuth indicator setting before each round is fired and use this as a deflection reference.

d. Adjustment should proceed rapidly until the center of impact is on the target. The enemy may use smoke to obscure the target, rendering further adjustment impossible. *Sufficient data must be obtained and recorded from the adjustment to allow effective fire to continue without observation.*

146. ATTACK BY INDIRECT LAYING METHODS. **a.** The procedure is quite similar to that described in part three. Guns are laid with gunner's quadrant and panoramic sight. Mechanical deficiencies of the azimuth indicator do not ordinarily permit the accuracy required by this type of firing.

b. Either a single gun or a platoon may be employed in this firing. When more than one gun is used, guns must be calibrated. (See FM 6–40.) Since the fire of a single gun can be more effectively controlled, the other gun or guns may follow the commands and be prepared to reinforce or replace the fire of the gun actually firing.

c. In determining the minimum elevation (par. 157), calculations must be made to fractions of a mil. It will frequently be necessary to omit the allowance for friendly troops on the mask to deliver fire on the target. In this event, the mask must be cleared of personnel or other precautions must be taken to protect the troops.

d. Gunners must lay to the nearest tenth of a mil for elevation, and to the nearest 0.5 of a mil in deflection. Care must be taken to eliminate all lost motion in the elevating and traversing mechanisms.

e. Fire is adjusted by precision methods. (See FM 6–40.) Fire should be conducted under the surveillance of a forward observer who can determine the effect and control the fire. In precision firing of this type, it is desirable for the observer to get as close to the target as possible.

f. A modified forward observation method may be used when fire is adjusted on openings in a hillside or on a cave with a vertical profile of 45° or more. The observer should be as close as possible to the target so he can estimate the points of impact to the nearest yard or less. The observer will sense each round in an estimated vertical plane; for example, "Two right, four above." Sensings are converted to mils at the gun position by reference to a mil conversion chart similar to the one shown below. (See fig. 77.) Deflection changes to the nearest $\frac{1}{2}$ mil and elevation changes to the nearest $\frac{1}{10}$ mil are made. When fire is adjusted on more level ground, the regular method of forward observation is employed.

MIL CONVERSION CHART

Range (yards)	Deviation in yards											
	⅓	⅔	1	2	3	4	5	6	7	8	9	10
	Correction in mils											
0												
100	3	7	10	20	30	40	50	60	70	80	90	100
200	2	3	5	10	15	20	25	30	35	40	45	50
300	1	2	3	7	10	13	17	20	23	27	30	33
400	.8	2	3	5	8	10	13	15	18	20	23	25
500	.7	1	2	4	6	8	10	12	14	16	18	20
600	.6	1	2	3	5	7	8	10	12	13	15	17
700	.5	1	1	3	4	6	7	9	10	11	13	14
800	.4	.8	1	3	4	5	6	8	9	10	11	13
900	.4	.7	1	2	3	4	6	7	8	9	10	11
1,000	.3	.7	1	2	3	4	5	6	7	8	9	10
1,100	.3	.6	.9	2	3	4	5	5	6	7	8	9
1,200	.3	.5	.8	2	3	3	4	5	6	7	8	8
1,300	.3	.5	.8	2	2	3	4	5	5	6	7	8
1,400	.2	.5	.7	1	2	3	4	4	5	6	6	7
1,500	.2	.4	.7	1	2	3	3	4	5	5	6	7

Figure 77. Chart for converting horizontal and vertical deviations in yards to deflection and elevation in mils for fire against vertical targets.

g. Dispersion must be considered. A single hit on the target is not necessarily sufficient evidence of properly adjusted elevation. Sufficient hits must be obtained to give a reasonable basis for assuming that the center of impact is on or near the center of the target.

147. TRANSFERS OF FIRE ON POINT TARGETS. a. General. When several targets have been located and identified, it is frequently advantageous to establish a firing chart based on survey and to fire successive transfers of fire on the several targets. The procedure is similar to that discussed in part four, except for the degree of accuracy required.

b. Survey. The locations of guns and targets are determined by survey and plotted on a 1/4000 survey chart. Surveys of position area and target area must be tied together. This may be done by means of a connecting survey or by using the same base for determining location of targets and guns. Coordinates of the starting point for the survey may be known or arbitrarily assigned. Similarly, the azimuth of the line used to determine direction may be known or arbitrarily assigned.

c. Angle of site. The angle of site (ch. 4) should be determined to the nearest tenth of a mil. It should include compensation for the height of the gun trunnions, where necessary.

d. Calibration. All guns should either be calibrated carefully or registered on a common base point.

e. Initial data. Each gun must be plotted on the firing chart. Deflection difference for a converged sheaf is not sufficiently accurate because of the marked change in the obliquity factor as fire is shifted across the sector. To converge the sheaf, individual deflection shifts must be computed for each gun. The position correction grid (FM 6–40) may be used for this purpose. Deflection shifts should be determined by using the coordinates of gun and target to compute the azimuth of the gun-target line. This azimuth, compared with the azimuth of the base line, will permit exact calculation of the necessary shift. Accurate range may be measured from the firing chart.

f. Corrections. Range corrections are computed on the basis of a precision registration on a base point. Base deflection is recorded after registration so that no deflection correction is necessary unless targets are widely separated in range.

g. Adjustment. When transfers of fire are computed as outlined in a through f above, first round hits will often be obtained. Even though a first round hit is not obtained, the time saved in adjustment justifies the time and energy expended in calculating the data. Adjustment should continue during fire for effect.

148. AMMUNITION. a. Adjustment. (1) HE with fuze, quick, is normally used during adjustment on hillside targets.

(2) When targets are on flat terrain, APC with base detonating fuze may be used at short ranges against targets of considerable profile. In the latter instance, the tracer element aids in sensing.

b. Fire for effect. The type of ammunition used

in fire for effect will vary with the type of target engaged.

(1) APC with base detonating fuze is used against caves in hard rock and against concrete structures to obtain an initial opening. When caving action is desired, this is followed by HE with fuze, delay, or fuze, CP, placed in the same opening.

(2) Against caves in medium rock, either the

76-mm GUN

Range (yards)	To shift from APC to HE subtract (mils)
500	0.2
600	0.2
700	0.3
800	0.3
900	0.3
1,000	0.4
1,100	0.4
1,200	0.5
1,300	0.5
1,400	0.6
1,500	0.6
1,600	0.7
1,700	0.8
1,800	0.9
1,900	1.0
2,000	1.0
2,100	1.0
2,200	1.1
2,300	1.2
2,400	1.2
2,500	1.2

Note. For Shell, HE, M42A1, and Shell, APC, M62.

Figure 78. Table for changing from APC to HE.

procedures described in (1) above, or the use of HE with fuze, CP, only may be effective.

(3) HE is used with fuze, M78, against caves in

soft rock or at long ranges against structures made of concrete.

(4) HE is used with fuze delay against caves in clay and soft earth.

(5) A cave or pillbox may be effectively neutralized by firing white phosphorus shell into the entrance or aperture.

c. Changing ammunition. In this type of firing it is often necessary to change the type of ammunition. The change is expedited by the use of a table such as figure 78. The table shown is designed for the 76-mm gun when changing from APC to HE. Similar tables can be prepared for any gun and any type of ammunition by comparing firing table elevations. In preparing such tables, interpolation is necessary to the nearest tenth of a mil.

Section V. FIRING WHILE MOVING

149. EMPLOYMENT. a. Cannon equipped with a gyrostabilizer can fire effectively from a moving vehicle. The caliber .30 coaxial, bow, and other machine guns can also be fired effectively while the vehicle is moving. The cannon should be fired from a moving vehicle only when absolutely necessary. The primary weapon in firing when moving is the caliber .30 coaxial machine gun. Proper use of the gyrostabilizer places effective fire on targets which otherwise could not be neutralized. *At any ranges where either tracer or strike can be observed, the coaxial machine gun is very effective when fired with the gyrostabilizer.*

b. In combat, the gyrostabilizer should be in operation whenever the vehicle is moving. It gives the gunner a relatively steady field of view, enables him to keep himself oriented with respect to the terrain, and allows him to search the terrain for targets.

c. In the final assault, both the psychological and the destructive effect are increased by the combined use of the coaxial and bow machine guns and cannon firing HE. The additional fire placed on the position increases the probability of destruction or neutralization of antitank weapons and reduces the losses of the attacking unit.

d. The cannon is used against surprise tank and antitank targets, particularly when moving down roads or over smooth terrain. At 500 yards, there is a 50 percent chance of obtaining a hit against a tank on the first round. This type of fire is not profitable at ranges over 1,000 yards.

e. The coaxial machine gun is used for reconnaissance by fire while moving. Suspected bushes, hedges, haystacks, and other similar targets should be engaged by fire. The cannon may be used against suspected buildings.

f. The gyrostabilizer is valuable in firing from landing craft and from amphibious tanks in assaults on beaches.

150. CREW TEAMWORK. **a.** Firing while moving requires close teamwork by the driver, gunner, and crew commander. The vehicle is driven in as nearly straight a line as possible. No rapid or jerky changes in direction which are not absolutely essential should occur while the gunner is laying the gun. *As soon as the gun is fired,* the driver corrects for drift of the vehicle. This is done quickly to prevent interfering with observation and laying for the next round.

b. The driver warns the gunner when rough terrain is ahead. When crossing rough terrain, the gunner does not fight the gun (attempting to keep it on the target by spinning the elevating handwheel), but waits until the stabilizer has regained control of the gun and the action has smoothed out.

c. The stabilizer does not lay the gun. It tends to keep the gun where it has been laid; that is, it eliminates extremely jerky vertical movements caused by the movement of the vehicle. Even with a stabilizer, the gun does not hold constantly on the target. The gunner watches the swing of the gun through the target and fires as the proper sight picture crosses the target.

d. Laying for deflection is easier when the vehicle travels at a slight angle to the target than it is when the vehicle is traveling straight toward the target.

151. STATIONARY TARGETS. a. Firing to front or rear. When firing to the front or rear, the gunner fires when the vertical line of the sight is on the target and as the proper range line crosses the top of the target if the gun is moving downward or as it crosses the bottom of the target if the gun is moving upward. If the upward and downward movements are so small that the proper range line does not fall below the bottom or above the top of the target, he fires as the proper range line crosses the center of the target.

b. Firing to side. When firing over the corner of the vehicle or to the side, the gunner uses a reverse lead to compensate for the vehicle's movement at an angle to the target. (See fig. 79.) The direction of the lead is opposite from that used in firing from a stationary vehicle at a moving target.

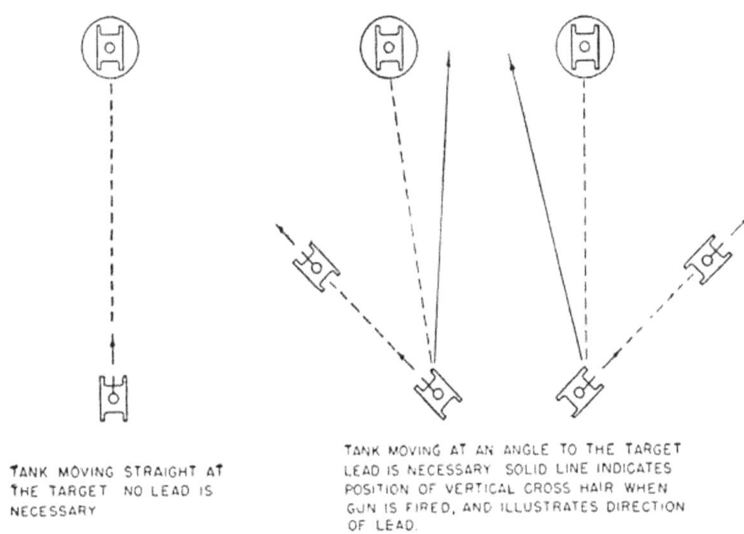

Figure 79. Firing from side.

When firing at a stationary target from a moving vehicle, the gunner establishes the lead on the side of the target opposite from the direction in which his vehicle is moving.

152. MOVING TARGETS. If the target and the vehicle firing are both moving, fire upon the target is not as accurate as when the vehicle is halted.

The vehicle is brought to a halt, the target is destroyed and then the advance is continued. In the final assault with the gyrostabilizer turned on, fire at moving targets is warranted.

PART THREE
INDIRECT FIRE BY PLATOON

CHAPTER 15

GENERAL

153. INTRODUCTION. **a.** Indirect fire is fire in which the gun is aimed for direction and elevation by means other than sighting at the target itself. There are two types of indirect fire, observed and unobserved. Part Three will deal with observed fire of the platoon. Part four covers the employment of the company as reinforcing artillery in the delivery of prearranged and unobserved fires.

b. Before training crews, the employment of indirect fire should be explained. Crew drill with simulated firing exercises should be conducted. After the men understand the procedure, simulated firing from tactical positions should be practiced. Terrain walks, with the crews selecting defiladed positions, should be conducted and their selections critiqued, keeping in mind the problem of minimum elevation. (See par. 156.)

154. TYPES OF MISSIONS. Two types of missions are fired from a defiladed position:

a. Fire for destruction. This is fire designed to destroy a point target such as an antitank gun, machine gun emplacement, or pillbox. Certain phases of this type of fire are discussed in section IV, chapter 14.

b. Fire for neutralization. This is fire designed to make an enemy's position untenable during the shelling, and to cause casualties and damages which will reduce the enemy's combat efficiency.

155. EMPLOYMENT. There are three methods of employing tank and tank destroyer guns in indirect fire.

a. Single gun in defilade. The gun is placed in a defiladed position. The crew commander places himself where he can observe the target. He determines the necessary data for the gunner to lay the gun, and he adjusts fire on the target. Communication between the crew commander and the gunner is ordinarily by voice or interphone extension. The procedure is discussed in chapter 16.

b. Two or more guns in defilade. (1) Where two or more guns are placed in a defiladed position, each crew commander may lay his own gun individually as in a above. It is more usual, however, to lay them parallel. One observer, usually the platoon leader or platoon sergeant, adjusts fire of all guns. Communication may be by wire, voice relay, or radio. The procedure is discussed in chapter 17.

(2) A platoon leader may use this method to establish a base of fire with part of his platoon while the remainder of the platoon maneuvers. A company commander may use it to establish

a base of fire with one platoon while the remainder of the company maneuvers.

c. Employment as reinforcing artillery. The most advanced type of indirect laying for tank and tank destroyer guns occurs when they are employed as reinforcing artillery. Employment in this role is discussed in Part Four.

156. MINIMUM ELEVATION. a. Whether firing a single gun or a platoon from a defiladed position, it is necessary to determine the minimum elevation. After boresighting, the gunner elevates the gun until the zero range line of the coaxial telescope clears the crest or mask. An alternate method is to sight along the bottom of the bore, and elevate the tube until the line of sight clears the mask. The elevation of the tube is measured with the elevation quadrant or the gunner's quadrant, and is reported to the crew commander. This angle is referred to as the "site to the mask."

b. For a single gun in defilade, the crew commander completes the computation of the minimum elevation. It is accomplished in the following manner:

(1) Add to the "site to the mask," the firing table elevation for the ammunition to be used at the range to the mask.

(2) Add two "forks" or two "c's" for the range to the mask.

(3) When the mask is occupied by friendly troops, add the angle subtended by 5 yards at the range to the mask. With high velocity guns, the decision to add this factor must be made after

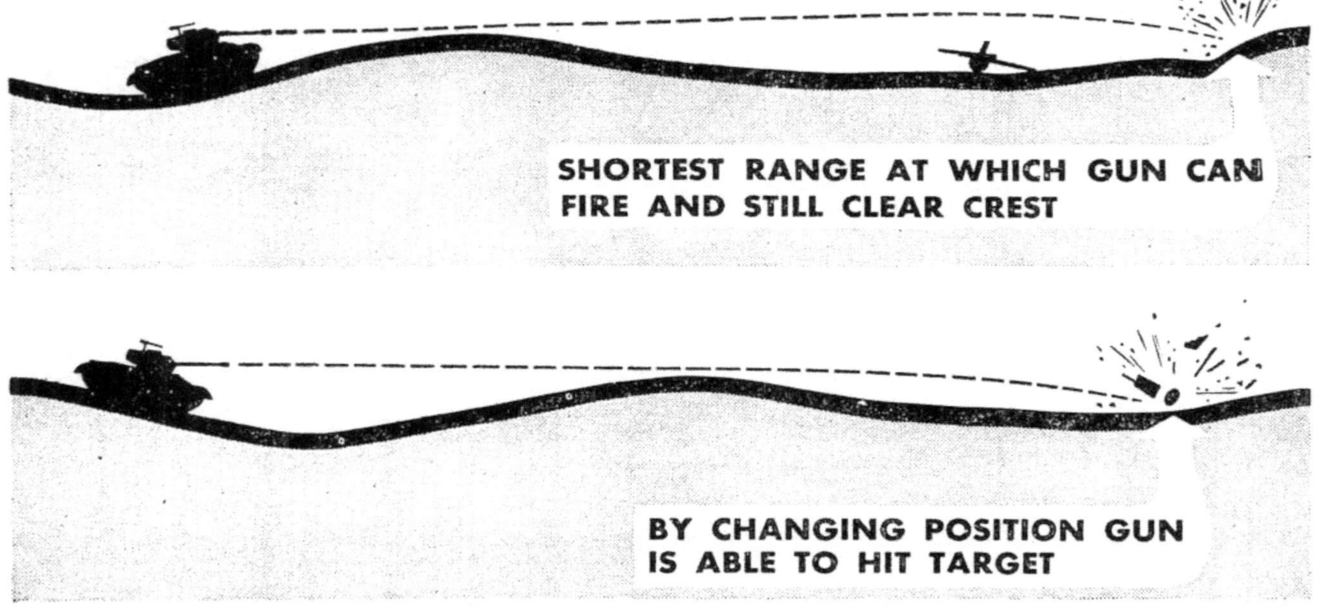

Figure 80. Placing guns for indirect fire.

careful consideration. It is not added arbitrarily.

(4) The sum is the minimum elevation. If it is fractional, use the next higher whole mil.

c. When firing a platoon from a defiladed position, the crew commander reports the site to the mask to the position commander, who completes the computation.

(1) The greatest "site to the mask" is used to compute the minimum elevation for the platoon as a whole.

(2) In some instances, a single narrow obstruction, such as a tree, may mask a single gun. Such an obstruction is not considered in determining the minimum elevation. If the gun cannot fire safely in a certain sector, it is called OUT and does not fire until called IN.

d. The position for the guns must be selected carefully. Defilade is desirable, but it must not be allowed to interfere with ability to accomplish the mission. In selecting the position, the crew commander must be certain that the minimum elevation will not prevent the gun from hitting the target. (See fig. 80.)

157. NUMBERS. To facilitate adjusting fire, guns are numbered, starting from the right flank, as No. 1, No. 2, No. 3, and No. 4. Thus, when using indirect laying, the gun on the right as one faces in the direction of fire is always designated No. 1, regardless of its permanent number in the platoon.

CHAPTER 16

SINGLE GUN IN DEFILADE

158. LAYING FOR DIRECTION. When firing a single gun, a common method of laying is for the crew commander to line in his gun and the target. He places himself on the line between the gun and the target and commands: LAY ON ME. The gunner lays on the crew commander and sets the azimuth indicator at zero. Any variation of the alidade method of laying mortars may be used as a refinement of this method of laying. (See FM 23–90.)

159. LAYING FOR RANGE. a. The crew commander determines the range from the gun to the target and estimates or measures the angle of site. (See ch. 4.)

b. If the commander has a firing table, he may convert the range into mils of elevation, apply a plus or minus angle of site, and announce the quadrant elevation.

c. If the crew commander does not have a firing table, he announces the range in yards and the angle of site in mils. The gunner refers to the aiming data chart to determine the elevation and applies a plus or minus angle of site to the firing table elevation.

d. The piece is laid with the elevation quadrant or gunner's quadrant.

e. The gunner may lay for range with the elevation quadrant and set off the angle of site with the graduated handwheel.

160. EXAMPLES OF FIRE ORDERS. a. An antitank gun is at 3,000 yards. The angle of site is zero. The crew commander is lining in the gun and the target.

GUNNER.
HE.
DELAY (QUICK).
LAY ON ME.
ANTITANK.
THREE THOUSAND (or QUADRANT ______).
FIRE.

b. An antitank gun is at 2,500 yards. The angle of site is minus 5 mils. The reference point (aiming point) is a lone tree on the sky line to the right front, 150 mils left of the target. Matériel: 75-mm gun, M3, HE M48.

GUNNER.
HE.
DELAY (QUICK).
REFERENCE POINT, RIGHT FRONT, LONE TREE ON SKY LINE.
RIGHT ONE FIVE ZERO.
ANTITANK.
TWO FIVE HUNDRED DOWN FIVE. (or QUADRANT THREE FIVE.)
FIRE.

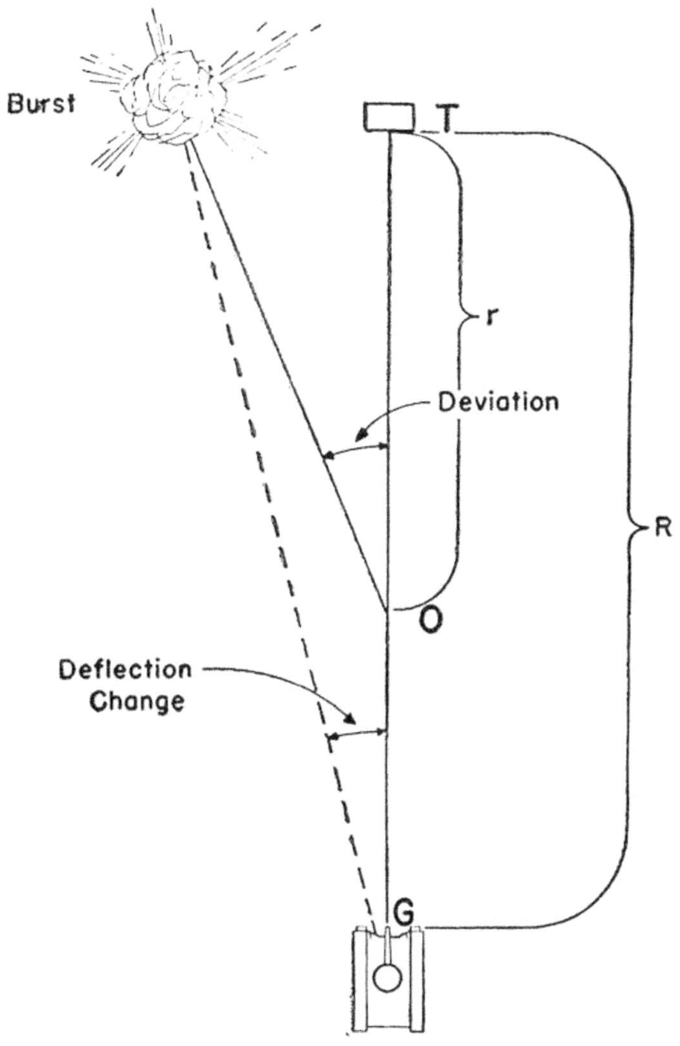

$$\frac{OT}{GT} = \frac{r}{R} = \frac{\text{Deflection Change}}{\text{Deviation}}$$

Figure 81. Comparison of observed deviation with deflection change.

161. RANGE AND DEFLECTION CHANGES. **a.** Range changes are made with the graduated handwheel or by changes of the setting on the elevation quadrant or gunner's quadrant.

b. Deflection changes are made with the azimuth indicator or panoramic sight.

162. ADJUSTING FIRE. **a.** The observer should stay as close to the line of fire as the terrain permits. Moving to the flank distorts sensings of range and deflection errors.

b. The observer adjusts fire upon the target as in direct fire. When the range from the observer to the target differs materially from the range from the gun to the target, the observed deviation differs from the actual deflection error. (See fig. 81.) The proper deflection correction is obtained by multiplying the measured deviation by the factor r/R. This relation is the observer-target range divided by the gun-target range.

CHAPTER 17

PLATOON IN DEFILADE

Section I. ORGANIZATION

163. GENERAL. When the platoon is employed in defilade, fire is more effective if control is centralized. Centralized control is obtained by having one person give the fire commands to the platoon. This individual is called the "position commander." Normally, he is the platoon sergeant. However, he may be the platoon leader. The term, "firing platoon," as used in this part of the manual, includes only the guns and personnel at the firing position when the platoon is prepared for action.

164. POSTS AND DUTIES OF INDIVIDUALS. a. Position commander. The post of the position commander is where he can best command the platoon. It is normally on the ground behind the center of the platoon. He is responsible for proper compliance with all orders and commands received at the position.

b. Telephone or radio operator. The telephone or radio operator is posted near the position commander. He transmits orders and commands to the position commander. When so ordered, he may transmit commands directly to the individual guns.

c. Recorder. The recorder is the chief assistant to the position commander and should always be posted near him. The duties of the recorder are discussed in paragraphs 167 and 168.

d. Others. When available, other assistants to the position commander should be used. These are mechanics, instrument personnel, sentinels, and noncommissioned officers in charge of ammunition. They are posted and perform their duties as directed by the position commander.

165. FIRING POSITION. a. Normally, there will be four guns at the position.

b. To centralize the control of the platoon, all commands are channelized through the position commander. Where possible, wire communication is established between the observation post and the platoon position. If wire communication cannot be established, it is usual for all the radios in the platoon to be turned off except one which becomes the fire control set. It is desirable to lengthen the cord to the microphone and headset on this control radio to permit its operation from the ground. This tends to keep control in the hands of the position commander by eliminating the necessity for shouting commands to him.

c. In transmitting commands to the individual guns, the position commander uses wire, voice, or voice relay. Radio is seldom used because of the fact that there is no available channel for this purpose. In a special situation, radio might be used, provided higher headquarters could assign a channel.

Section II. SELECTION AND OCCUPATION OF POSITION

166. GENERAL. **a.** In selecting gun positions, these factors should be considered:

(1) Field of fire.

(2) Defilade from enemy observation.

(3) Cover and concealment.

(4) The placing of the guns so that, when they are laid parallel, the sheaf (sec. VII, ch. 17) will cover an area three times the effective width of burst. The effective width of burst of a 75-mm shell is 40 yards; of a 76-mm shell, 30 yards; of a 90-mm shell, 40 yards; of a 105-mm shell, 50 yards. For example, with defilade, concealment, and a good field of fire, the ideal width of the platoon front would be 100 yards for the 75-mm gun.

b. The position should be occupied promptly, smoothly, and with a minimum of confusion. Every effort is made to maintain and to take advantage of concealment. No indications should be given of the presence of a platoon. New tracks are obliterated. Surplus vehicles are sent away from the position.

c. The sequence of operations in preparing for fire from defiladed positions is designed to cut the time element of this operation to the minimum. All of these operations must be performed if accuracy and safety are to be insured. Preparing for action starts when the platoon arrives in the firing position, but many of the operations may be facilitated by careful checks and adjustments before the platoon leaves the assembly or bivouac area. Steps in preparation for firing from defiladed positions are as follows:

(1) The gunner's quadrant is tested, and the elevation quadrant is tested and adjusted.

(2) The coaxial telescope (panoramic sight) is boresighted.

(3) The azimuth indicator is tested.

(4) A report is made to the position commander when the gun is ready to be laid.

(5) For range firing the safety stakes are established by the safety officer. (See app. I.)

(6) The "site to the mask" for the sector is measured and reported to the position commander.

(7) The ammunition is stowed in the vehicle.

(8) A report is made to the position commander when the gun is ready to fire.

167. DUTIES OF PERSONNEL IN PREPARING FOR FIRING. a. General. These duties are listed here as a guide. Individual duties prescribed in pertinent Field Manuals will determine the exact functions of each crew member.

b. Position commander. (1) Supervises the occupation of the position.

(2) Determines whether any gun will have a minimum elevation so great as to prevent firing in the desired target area. (See fig. 80.)

(3) Supervises the preparation for firing by the individual guns.

(4) Lays the platoon. (See par. 175.) Normally, the platoon is laid initially in the approxi-

mate direction of fire on a definite Y-azimuth which is a multiple of 100 mils.

(5) Checks each gun carriage for cant.

(6) Computes the minimum elevation. (See par. 156.)

(7) Reports to the observation post—

(*a*) The platoon is ready.

(*b*) The minimum elevations, charge (____) (____).

(*c*) The width of the platoon front.

(*d*) Visible aiming points.

c. Recorder. (1) Sets up the aiming circle for the position commander.

(2) Keeps a record of the site to the mask reported by each gun.

(3) Records the deflections of the safety stakes announced by the safety officer and gives these deflections to any gun on call.

(4) Records the maximum and minimum elevations for each charge computed by the position commander for the platoon as a whole.

(5) Records all messages and commands that come from the OP.

d. Telephone or radio operator. (1) Checks to see that all radios are turned off, except as noted in paragraph 165.

(2) When wire is used, insures that there is enough slack in the telephone line to allow him to follow the position commander anywhere in the position area.

(3) Transmits all commands and messages directly to the position commander and to the recorder.

e. Crew commander. (1) Puts the gun into as nearly level a position as possible under the supervision of the position commander.

(2) Puts the cross hairs in the witness lines on the muzzle.

(3) Wipes off the front of the coaxial telescope.

(4) Checks gun for cant with quadrant, M1.

(5) Reports to the position commander when his gun is ready to be laid, and repeats commands to the gunner when necessary.

f. Gunner. (1) Assisted by the loader, tests the gunner's quadrant, and tests and adjusts the elevation quadrant.

(2) Assisted by the loader and crew commander, boresights the coaxial telescope or panoramic sight.

(3) Tests the azimuth indicator.

(4) Reports to the crew commander that his gun is ready to be laid.

(5) Lines in the safety stakes after the gun has been laid, if required.

(6) Determines the site to the highest part of the mask and reports it to the crew commander.

(7) Reports to the crew commander that he is ready, and assists the loader in completing the stowage of ammunition. (If safety stops are being used, the gunner will place the safety stop as soon as the minimum elevation is announced.)

g. Loader. (1) Assists the gunner in testing the gunner's quadrant and in testing and adjusting the elevation quadrant.

(2) Assists the gunner in boresighting.

(3) Stows the ammunition in the turret.

h. Driver. (1) Selects ground as nearly level as possible.

(2) Prepares the ammunition for firing, removing it from the cases.

(3) Sets out the aiming posts and safety stakes under the direction of the gunner and crew commander.

Section III. DUTIES OF PERSONNEL DURING FIRING

168. GENERAL. a. The duties as listed herein are intended to serve as a guide when no other reference is available. The detailed lists of duties prescribed in pertinent field manuals should be used wherever possible.

b. *The position commander* controls the platoon. He is responsible for the execution of each command that comes to the firing platoon. He performs the following duties:

(1) Checks the deflection and elevation of each gun periodically to insure that commands have been executed accurately.

(2) Gives all commands to the firing platoon.

(3) Supervises the execution of each command by the guns, checking on the cause of delays and the manner in which the members of the platoon perform their duties.

(4) Checks on ammunition supply and the distribution of ammunition within the platoon.

c. *The recorder* is the chief assistant to the position commander. He performs the following duties:

(1) Keeps a record of all messages and commands received at the firing position.

(2) Keeps a record of the deflection and elevation settings for each gun, and is prepared to give the proper settings to the position commander or a crew commander on call. (This information is entered on a Recorder's Sheet, as shown in FM 6–40.)

(3) Keeps a record of the data sent to the platoon on the data sheets. (See FM 6–40.)

(4) Reports expenditure of ammunition and the amount remaining in the platoon after each fire mission.

d. *The telephone operator* has the following duties:

(1) Follows the position commander, repeating fire commands and messages to him and to the recorder.

(2) Reports the firing of each round or series of rounds by the report ON THE WAY. It may be desirable to include, which gun or guns are firing, the number of rounds being fired, the quadrant or range which was fired, and when the number of rounds commanded have been completed.

e. *The crew commander* takes a position from which he can control the crew and observe the position commander. He performs the following duties:

(1) Repeats commands to the gunner if necessary.

(2) Repeats all of the gunner's messages to the position commander or the recorder.

(3) Supervises and checks on the work of the gunner and the remainder of the crew.

(4) Watches the safety stakes, if used, and reports unsafe to fire whenever his gun is pointed outside the safety limits.

f. *The gunner* executes the fire commands. He performs the following duties:

(1) Lays the gun for deflection.

(2) Lays the gun for elevation.

(3) Marks safety limits on azimuth indicator, as a guide.

(4) Fires the gun.

g. *The loader* has the same duties as in direct fire.

h. *The driver* performs the duties prescribed in the appropriate manuals on crew drill.

169. CHECKING SETTINGS. **a.** The position commander usually checks settings and layings only during lulls in firing. If he doubts the accuracy of the laying of any gun, he calls the gun out of action.

b. If the platoon is firing close to friendly troops, the position commander continually checks the laying.

170. REPORTING ERRORS. All errors reported to the position commander are corrected promptly and reported to the OP.

Section IV. FIRE COMMANDS AND THEIR EXECUTION

171. DEFINITIONS. **a.** *Fire commands* are commands which convey all the information necessary for the commencement, conduct, suspension, and cessation of fire, and activities incident thereto.

b. *Firing data* are the elements of fire commands which prescribe the settings for instruments, fuzes, elevation quadrants, or gunner's quadrants.

c. *The base piece* is the gun for which initial data are computed. These initial data are used to determine the data for the other guns.

d. *The base point* is a well-defined point, in the target area, whose location is known.

e. *The base line* is the line passing through the base piece and the base point. It is used as an origin for shifts.

f. *The base angle* is the horizontal clockwise angle from the base line to the orienting line. (See par. 194b.) It is never greater than 3,200 mils.

g. *The base deflection* is the deflection on the panoramic sight or azimuth indicator from which further shifts are made.

172. SEQUENCE. The sequence for transmission of fire commands, as given by the position commander, is listed below. For a discussion of the various elements see FM 6–40.

Pieces to follow commands.
Ammunition.

Charge.
Fuze.
Direction.
Distribution.
Site.
Pieces to fire.
Method of fire.
Use of quadrant, M1, or quadrant, M9.
Elevation.
Command to fire.

173. COMMAND TO FIRE. To follow direct-fire procedure tank and tank destroyer guns will use the command FIRE (ch. 8) for the last element of the fire commands as given at the gun position.

Section V. LAYING THE PLATOON

174. GENERAL. The position commander lays the platoon parallel initially and whenever he is ordered to record base deflection. Generally, initial laying is on a Y-azimuth (compass). It is accomplished by reciprocal laying.

175. RECIPROCAL LAYING. Reciprocal laying is laying a gun or fire control instrument for direction by means of another instrument or another gun. It is accomplished by making the 0–3200 line of the panoramic sight or azimuth indicator parallel to the 0–3200 line of the laying instrument. The position commander when using the aiming circle establishes the 0–3200 line of the instrument in the direction of fire, which may be a compass direction. The command to the position commander is: COMPASS (SO MUCH). The position commander does not repeat the command. The position commander sets up the aiming circle away from magnetic influences in a place where it can be used as an aiming point for all guns. (See fig. 82.) The position commander—

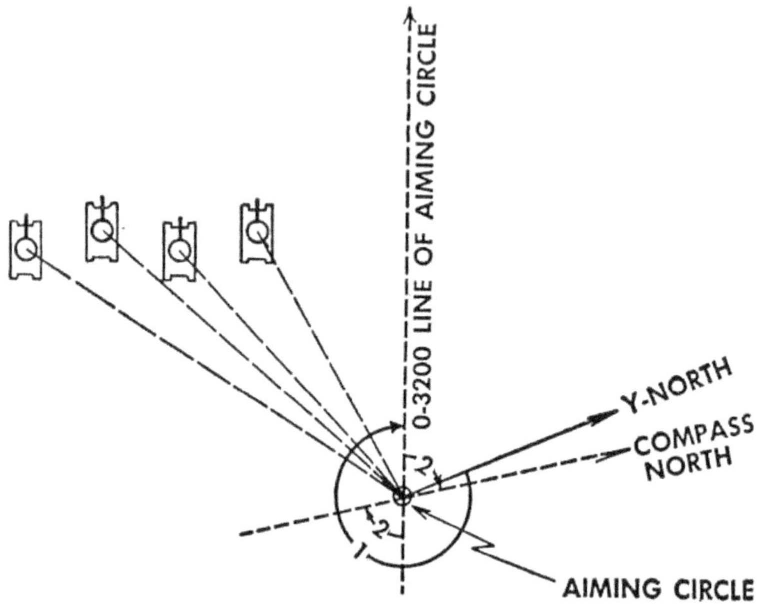

Figure 82. Laying by compass.

(1) Subtracts the announced Y-azimuth (compass) (angle 1) from the declination constant of the aiming circle (adding 6400 to the declination constant if necessary).

(2) Sets the remainder (angle 2) on the azimuth and micrometer scales of the aiming circle.

(3) Releases the compass needle and centers it with lower motion. (The 0–3200 line of the

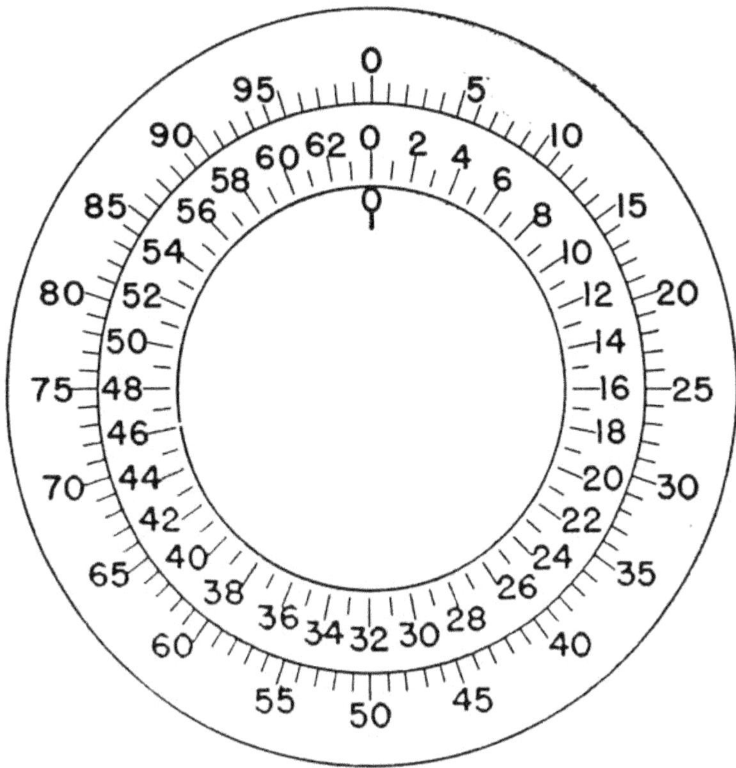

Figure 83. Dials of azimuth indicator, M19.

instrument now coincides with the announced Y-azimuth (compass).) He then lays the guns reciprocally. The matériel used will determine the procedure to be followed.

a. Laying by aiming circle. (1) *Panoramic sight.* See FM 6–40.

(2) *Azimuth indicator, M19 or M20.* The scales of the M19, M20, and M21 indicators are shown in figures 83 and 84.

(*a*) The position commander alerts the gun-

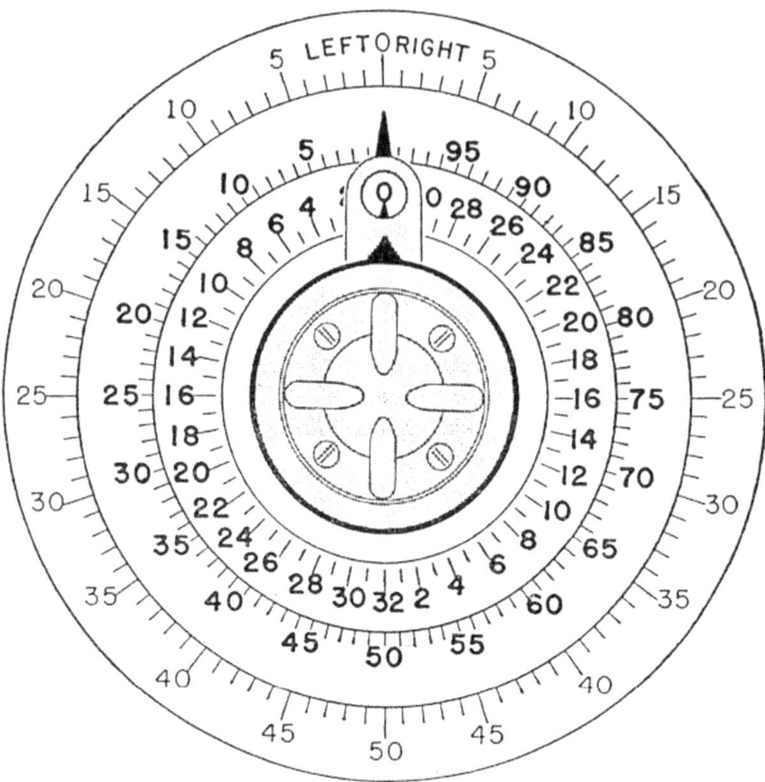

Figure 84. Dials of azimuth indicators, M20 and M21.

ners with the command AIMING POINT THIS INSTRUMENT.

(*b*) The gunners traverse onto the aiming circle and set their indicator scales at zero. The position commander refers to the coaxial telescope of each gun with the upper motion of the aiming circle and announces DEFLECTION NO. 1____, NO. 2____, NO. 3____, NO. 4____.

(*c*) The gunners traverse until the deflection reading announced by the position commander is on their azimuth indicators. Since the scale is

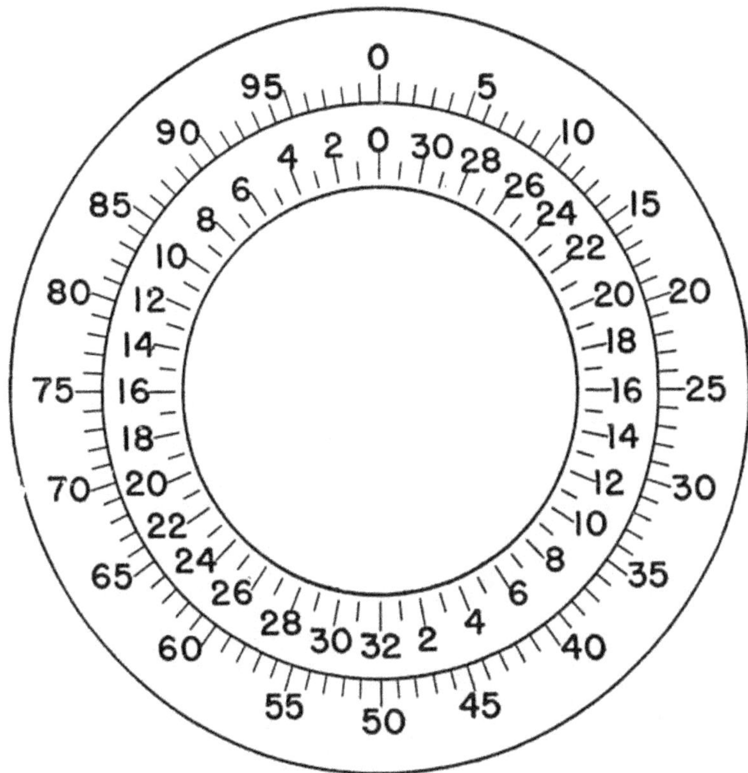

Figure 85. Dials of azimuth indicator, M18.

divided into two 0–3200 portions, the gunners must know the proper direction of fire so that they may use the appropriate portion of the scale.

(3) *Azimuth indicator, M18.* The scales of this indicator are shown in figure 85.

(*a*) The position commander alerts the gunners with the command AIMING POINT THIS INSTRUMENT.

(*b*) The gunners traverse onto the aiming circle and set their indicator scales at zero. The position commander refers to the coaxial telescope of each gun with the upper motion of the aiming circle and announces an individual shift for each gun. The rules for these shifts are as follows:

1. If the aiming circle is to the left of the line of fire of the piece, the command is RIGHT. Whether the aiming circle is to the front or to the rear of the piece, the amount of the shift is 3,200 minus the reading on the scales. (See a, fig. 86.)

2. If the aiming circle is to the right of the line of fire of the piece, the command is LEFT. Whether the aiming circle is to the front or to the rear of the piece, the amount of the shift is read directly from the scales. (See b, fig. 86.)

b. Laying parallel by another gun. (1) *Panoramic sight.* See FM 6–40.

(2) *Azimuth indicator M18, M19, M20, and M21.* (*a*) The position commander commands: ON NUMBER____, LAY PARALLEL.

(*b*) The gunners of the guns to be laid traverse onto the sight of the announced gun and zero their indicators.

(*c*) The gunner on the laying piece, having zeroed his azimuth indicator, traverses onto the sight of each gun in turn and announces the deflection. For the M18, M19, M20, and M21 indicators, this deflection is obtained by subtracting the reading on the indicator from 3,200. For the M18 indicator, the direction of traverse is also announced. It is the same as the direction the laying piece had to traverse. No direction need

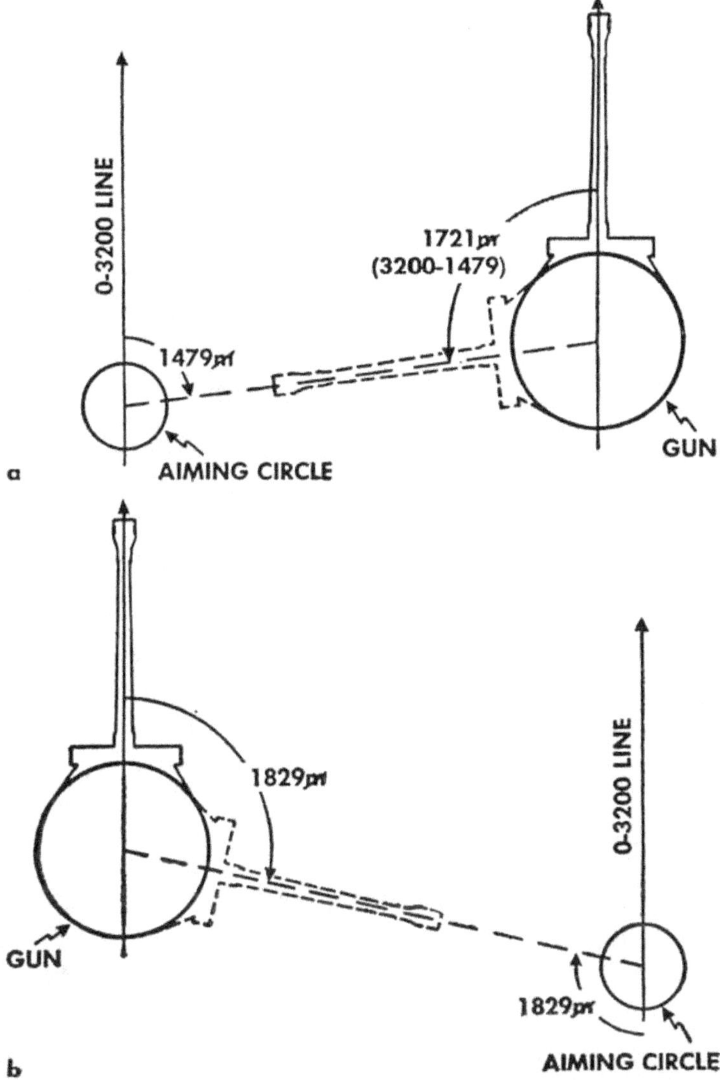

Figure 86. Reciprocal laying with M18 indicator.

be announced with the M19, M20, and M21 indicators.

176. RECORDING BASE DEFLECTION. **a.** The officer conducting fire may order base deflection recorded at any time.

b. The position commander may have base deflection recorded after the initial laying or after a change of aiming points.

c. The command is: RECORD BASE DEFLECTION. Without changing the laying of the gun, each gunner sets his azimuth indicator at zero and lays on aiming posts or an aiming point. With the panoramic sight, he refers to an aiming point or aiming posts.

d. Once base deflection has been recorded, it is not changed except at the command of the person conducting fire.

e. If the person conducting fire desires to have base deflection recorded with the pieces laid other than parallel, he adjusts the sheaf and then commands: AS LAID, RECORD BASE DEFLECTION.

177. USE OF AIMING POSTS. When recording base deflection, aiming posts are put out if practical. The command is: AIMING POINT, AIMING POSTS. With the gun on base deflection, the posts are lined in with the coaxial telescope or panoramic sight. The far aiming post is placed at least 100 yards from the gun, and the near post halfway between the gun and the far post. It is important that the near aiming stake be one-half the distance between the far stake and the gun. When it is not practicable to put out aiming posts, the gunner traverses (refers) to a prominent, readily identifiable aiming point, and notes the reading on his azimuth indicator (panoramic sight). The guns may be laid for direction on future targets by shifts right or left from base deflection.

178. CORRECTION FOR DISPLACEMENT. When the gunner notes that the vertical hair of the telescope is displaced from the line formed by the

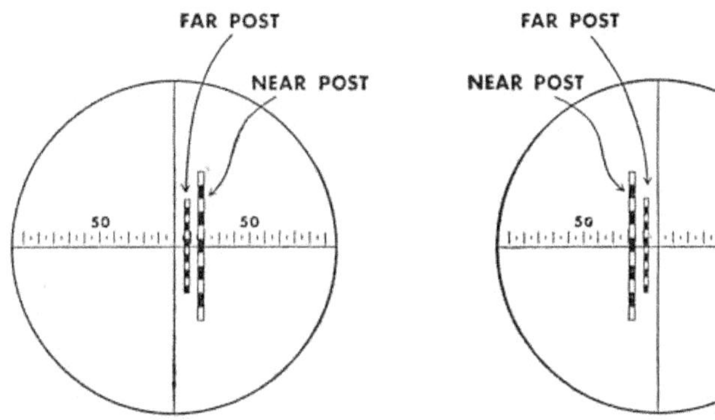

Figure 87. Correction for lateral displacement using the panoramic sight.

two aiming posts (or aiming post lights) he lays in such manner that the far aiming post (light) appears exactly midway between the near aiming post (light) and the vertical hair and proceeds as follows:

a. With panoramic sight. (1) If traverse of the piece causes displacement, the gunner continues to lay as described above.

(2) If shifts in position of the carriage cause excessive displacement the gunner will notify the crew commander. At the first lull in firing, the crew commander will notify the position commander and request permission to realine the aiming posts.

(3) When realinement is directed, the gunner—

(*a*) Lays with the sight picture described above. (See fig. 87.)

(*b*) Realines far aiming post.

(*c*) Realines near aiming post.

(4) If impracticable to move one of the aiming posts, the gunner—

(*a*) Lays with the sight picture described above.

(*b*) Refers to post which can not be moved.

(*c*) Realigns other post.

(*d*) Reports new deflection setting.

b. With coaxial telescope. (1) Zeros azimuth indicator. The gun is now laid parallel to the base line.

(2) If progressive shifts in position of the carriage cause displacement, the gunner will notify the crew commander. At the first lull in firing, the crew commander will notify the position commander and request permission to realign the aiming posts.

(3) When realinement is directed, the gunner—

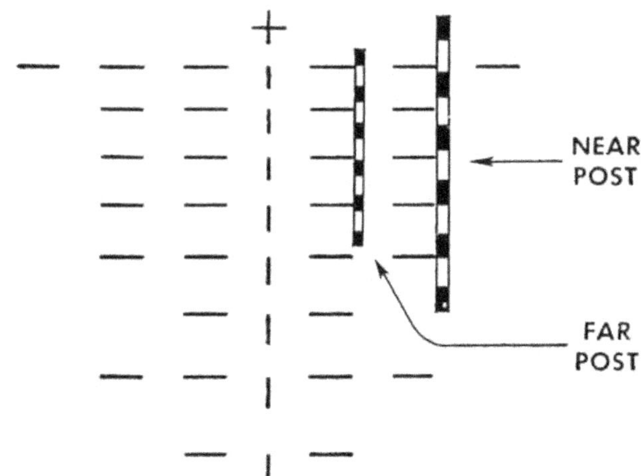

Appearance of aiming posts in reticle when telescope has displaced to LEFT of line of aiming posts

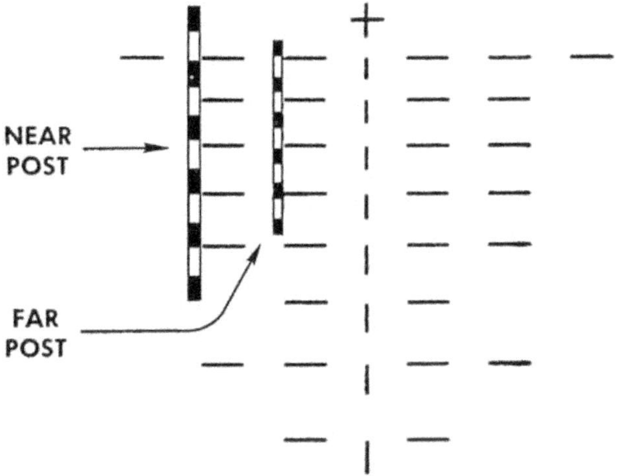

Appearance of aiming posts in reticle when telescope has displaced to RIGHT of line of aiming posts

Figure 88. Correction for lateral displacement using the coaxial telescope.

(*a*) Lays with the sight picture described above. (See fig. 88.)

(*b*) Zeros the azimuth indicator.

(*c*) Realines far post.

(*d*) Realines near post.

179. MEASURING COMPASS. The officer conducting fire may command MEASURE THE COMPASS. The position commander does not repeat the command. The procedure is as follows:

a. Panoramic sight. See FM 6–40.

b. Azimuth indicators, M18, M19, M20, and M21.

(1) The position commander sets up his aiming circle with the 0–3200 line in the approximate line of fire. (See fig. 89.)

(2) He commands: NUMBER____, AIMING POINT THIS INSTRUMENT, MEASURE THE DEFLECTION.

(3) The gunner zeros his indicator, traverses to the aiming circle, and reports his reading as NUMBER____, DEFLECTION____.

(4) With azimuth indicator, M18, the position commander uses this direct reading when the aiming circle is right of the line of fire. If left of the line of fire, he uses 3200 minus the reading. When the result is set on the aiming circle, he lays on the coaxial telescope, using the lower motion. With azimuth indicators, M19, M20, and M21, the position commander subtracts the reading from 3200, sets the result on the aiming circle, and lays on the coaxial telescope, using the lower motion.

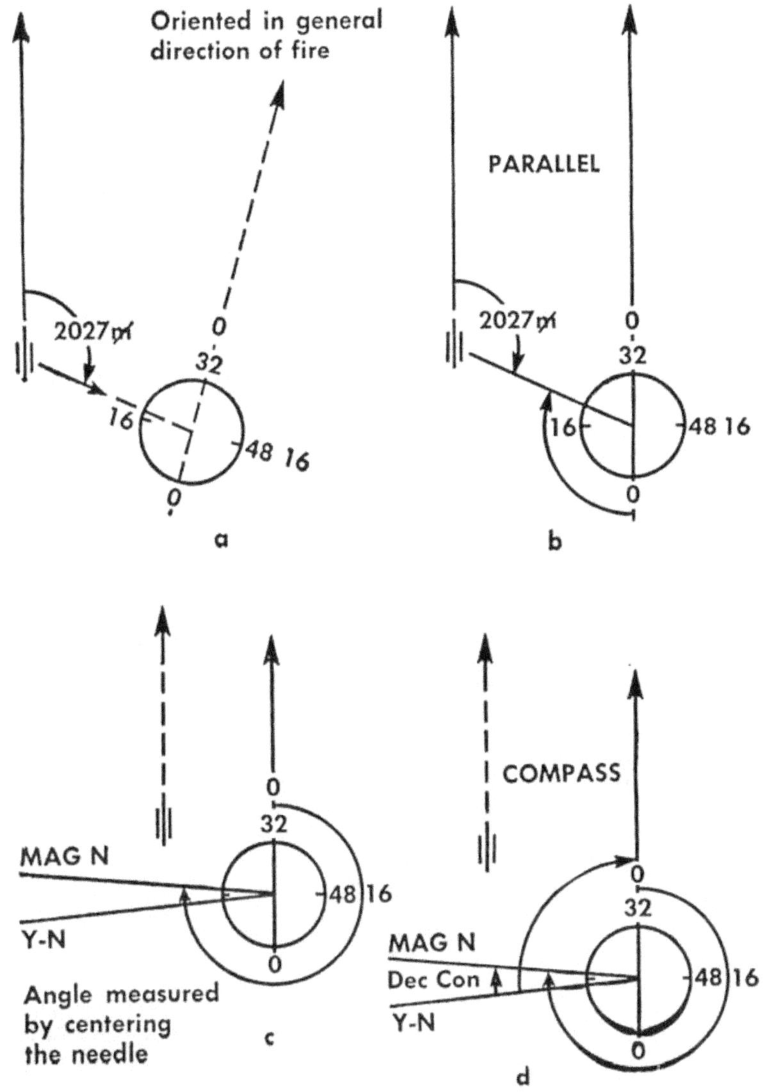

Figure 89. Measuring compass.

(5) He centers the needle, using the upper motion.

(6) He subtracts the reading on the scales from the declination constant (plus 6400 if necessary).

(7) He reports this as COMPASS ____.

180. REPORTING ADJUSTED COMPASS. a. With azimuth indicators, M18, M19, M20, and M21, the

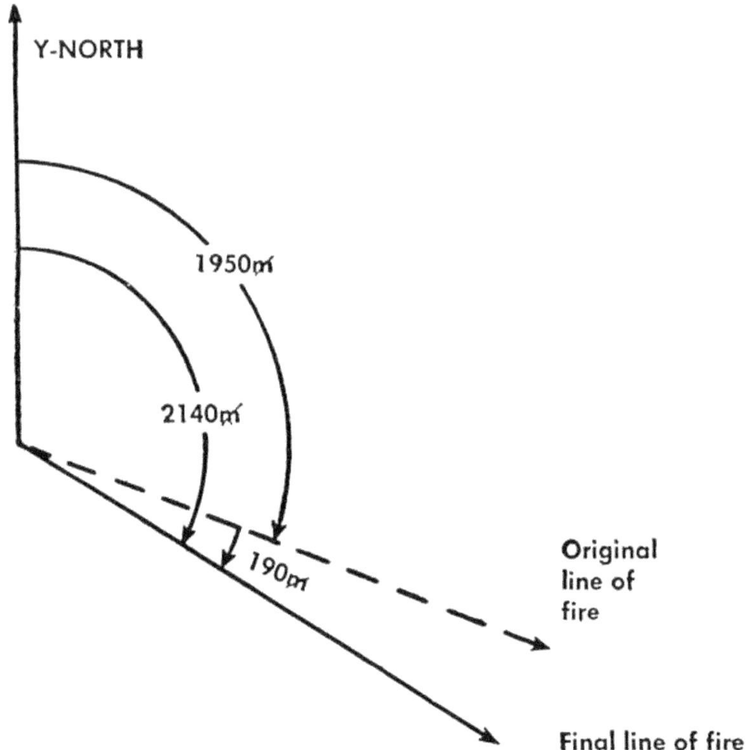

Figure 90. Adjusted compass.

position commander personally zeros the indicator and traverses the gun onto the aiming point. The amount that is traversed is the base deflection shift or the adjusted deflection. (When firing from the 76-mm Gun Motor Carriage, M18, the turret is traversed back to the azimuth indicator reading at which the gun was fired. This compensates for the error caused by turret twist which does not affect the flight of the projectile.) The base deflection shift is applied to the Y-azimuth on which the piece was laid previously when base deflection was recorded, and is reported as ADJUSTED COMPASS. (See fig. 90.)

b. For procedure with the panoramic sight, see FM 6–40.

Section VI. DETERMINATION OF DIRECTION

181. OBSERVATION POST METHOD. If the observer can see both the target and his platoon from his observation post, he can place the 0–3200 line of his aiming circle on the target and lay each gun individually.

a. If the observer is on the line of fire, laying the guns parallel lays them on the target.

b. If the observer is not on the line of fire, it is necessary to correct for the target offset. The procedure for determining direction with a target offset is described in FM 6–40. Size of the target offset may be computed by use of the mil formula.

Section VII. DISTRIBUTION

182. WIDTH OF SHEAF. Guns are normally laid with a parallel sheaf. The width of the sheaf is the lateral distance between flank bursts. The width of a parallel sheaf is thus the lateral distance between the flank guns of the platoon.

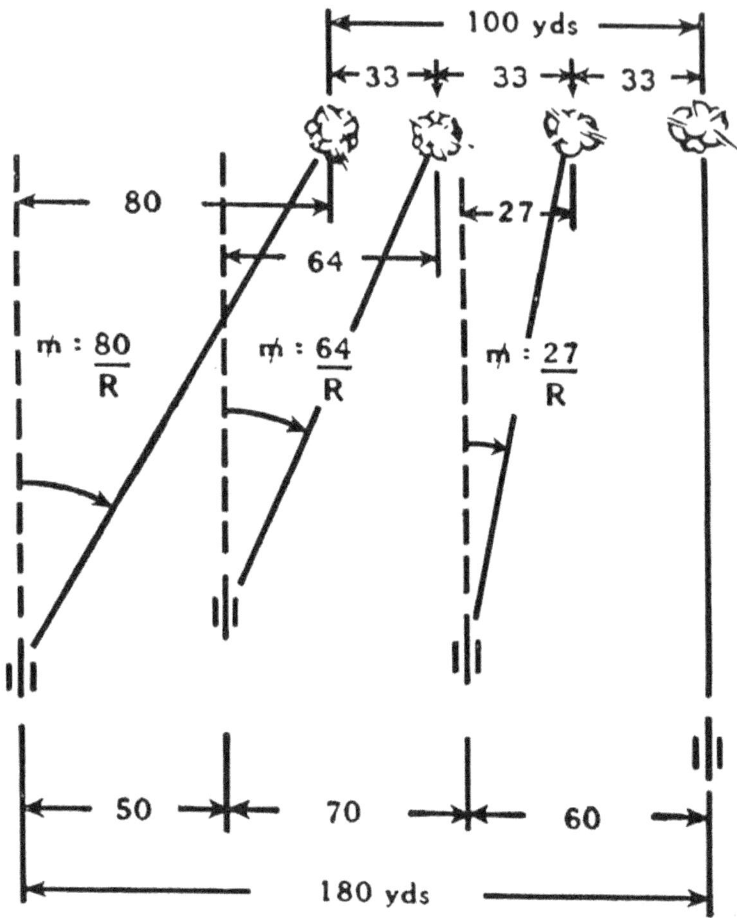

Figure 91. Correction of sheaf to platoon front.

183. CORRECTION OF SHEAF. **a.** Although guns may not be equally spaced, it is not usually necessary to make corrections. If it is necessary to

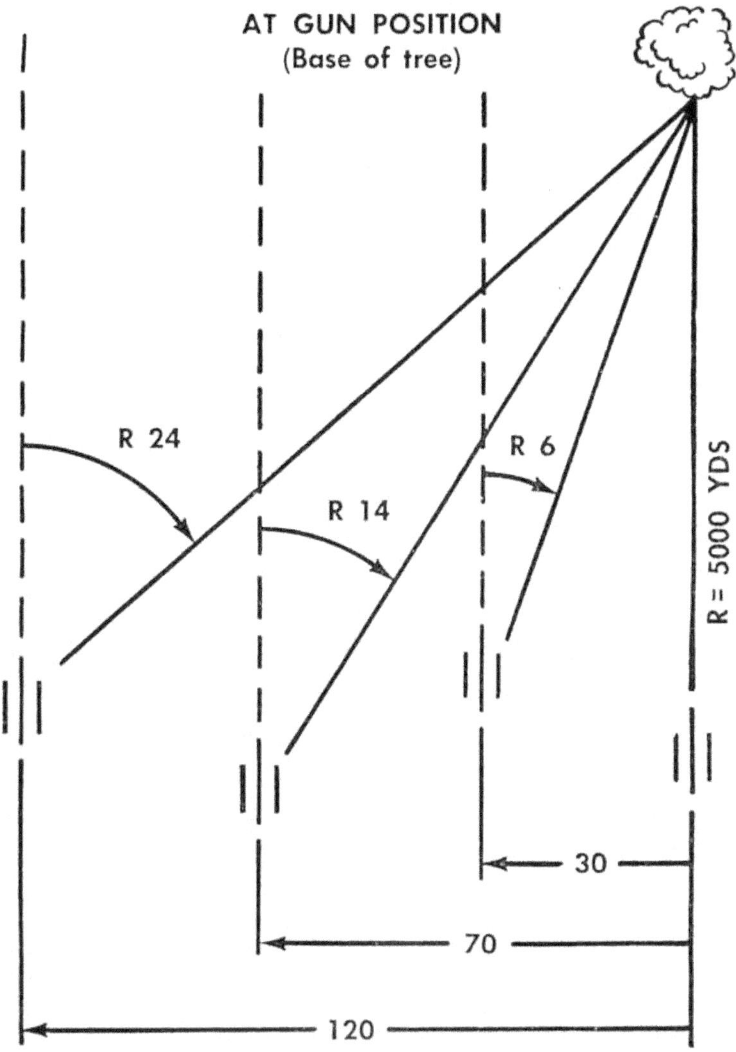

Figure 92. Converging sheaf.

correct the sheaf because of the platoon front, individual shifts are determined by the mil formula. This is computed (fig. 91) by using the distance between guns (W) and the range to the target

(R). To control the sheaf with a deflection difference command, see FM 6–40.

b. A point target, such as an antitank gun position, may require a converged sheaf. This is accomplished by determining individual shifts for each gun by the mil formula. (See fig. 92.)

c. A target whose front is greater than the width of the sheaf is neutralized by shifting the entire sheaf. This is accomplished by one command to the entire platoon rather than by individual shifts of each gun.

Section VIII. CONDUCT OF FIRE

184. GENERAL. To conduct the fire of two or more guns properly, the observer must consider the mission. When attacking an accurately located point target, fire is converged on that point. With an area target, the purpose is to lay down the fire of a platoon with such surprise and speed that the maximum destruction, demoralization, and casualties result. Accuracy is necessary, but speed is equally important.

185. ADJUSTMENT. **a.** Although the target may be an area, the observer ordinarily selects a specific adjusting point in or near the area for ease in sensing.

b. Adjustment is started with a parallel sheaf. Platoon salvos, beginning from the downwind flank, are used. If ammunition must be conserved, adjustment may be started with one or both of the interior pieces.

c. Fire against area targets may be controlled and adjusted by forward observation methods. This method is discussed in FM 6–135 and 6–40. Normally, altitude sensings are not employed.

d. In firing against point targets, precision methods of adjustment are used. (See FM 6–40 and paragraph 146, this manual.)

e. When time and training permit, the construction of an observed fire chart will enhance the effectiveness of this fire. The construction and use of this chart is described in FM 6–40. When the observed firing chart is used, altitude sensings may be employed.

PART FOUR

EMPLOYMENT AS REINFORCING ARTILLERY

CHAPTER 18

GENERAL

186. EMPLOYMENT. In certain situations tanks and tank destroyers may be required to fire as reinforcing artillery. FM 6–40 prescribes the technique to be employed. Succeeding paragraphs outline differences in technique occasioned by the nature of the matériel.

187. TRAINING. Training as reinforcing artillery is undertaken only after crews have attained a high standard of proficiency in the primary role.

188. TYPES OF FIRES. **a.** Artillery fire is classified as "observed" and "unobserved."

b. Observed fire is comparable to that described in Part Three.

c. The delivery of unobserved fires by tank and tank destroyer units is subject to limitations as follows:

(1) Corrections are based usually on registration. However, they may be based on meteorological data when available.

(2) Fire is normally delivered on prearranged targets.

(3) Responsibility for survey ends with the completion of a position area survey, which includes tying in with a place mark furnished by the artillery. Platoons may be laid by base angle.

(4) Fire at center range is normal. However, in some instances, zone fire may be appropriate.

(5) Firing chart will normally be a grid sheet. The type of firing chart will be prescribed by the reinforced field artillery.

189. RESPONSIBILITY. **a.** To facilitate execution of prearranged or unobserved fires, the reinforced field artillery has the following specific duties.

(1) Designation of position area, usually for each company. At times, designation of an area for an entire battalion or for an individual platoon may be proper.

(2) Designation of direction of fire and general location of the target area.

(3) Designation of at least one place mark location in each company area. The coordinates and altitude of the place mark and a line of known direction will also be furnished.

(4) Designation of a base point by location on the ground, and by furnishing its coordinates and altitude for plotting on the firing chart.

(5) Designation of targets, to include location, altitude, ammunition allotted, and the time of firing.

b. It is seldom practicable to have all of the above data available at the beginning of the preparation for unobserved fire. Successive operations are completed as data become available.

CHAPTER 19

PROCEDURE

Section I. SEQUENCE

190. GENERAL. The steps indicated in this paragraph are necessary to preparation of unobserved fires. The sequence may vary in some particulars but generally is as follows:

a. Selection of positions for individual platoons.

b. Location of a place mark by the field artillery.

c. Survey by the reinforcing company to—

(1) Locate the base piece of each platoon.

(2) Establish the orienting lines.

d. Designation of ground and firing chart location, and of the altitude of the base point by the field artillery.

e. Laying the platoons by base angle.

f. Designation by the field artillery of the coordinates and altitudes of concentrations, ammunition allotted, and time of firing.

g. Computation of data.

h. Registration.

i. Computation and application of corrections.

191. SELECTION OF POSITION FOR INDIVIDUAL PLATOONS. The position area for the company having been selected by the field artillery, the company commander locates the firing position for each platoon. This is facilitated if the direction of fire and the range to the target area are known, since the minimum elevation must be taken into account. Unless positions are to be occupied promptly, the positions of the base piece of each platoon are marked to facilitate survey. The presence of survey personnel during the selection of positions often prevents misunderstandings.

Section II. POSITION AREA SURVEY

192. GENERAL. a. The object of any survey is to locate various points in relation to each other, both horizontally and vertically. Surveying is map making. The only points required on a map for determination of firing data are the location of the base point, the base pieces of each platoon, and the targets upon which fire is to be placed. The base pieces of each platoon are the only points located by the reinforcing unit. The reinforced field artillery surveys the base point and the targets and reports their altitudes and their locations by coordinates or overlay.

b. To save time, survey is started by the company as soon as the platoon positions have been selected. If the place mark has not been staked, the traverse may start from any chosen point. The field artillery survey is brought to this point later, or a connecting traverse is run to the place mark.

c. The company commander designates a place where survey data are to be sent. As survey data are reported, they are plotted on the firing chart.

193. LOCATION OF PLACE MARK. **a.** The purpose of the place mark is to serve as a point of origin for the position area survey. For this reason it must give location, altitude, and direction. This information is written on the place mark stake or on an attached tag. (See fig. 93.)

(1) Location is specified by coordinates.

(2) Altitude is given in yards. It is either given as true altitude or with respect to an arbitrarily selected datum plane that is to be used.

(3) The method of establishing direction is to give the Y-azimuth from the place mark to a visible point. If this azimuth disagrees slightly with the azimuth as measured, it should be accepted nevertheless, since it will tie the position area survey to the artillery's target area survey.

b. The place mark is selected to provide a good starting point for the position area traverse. It may be on the orienting line.

c. The unit named on the place mark tag is the unit which established the place mark.

194. DETAILS OF SURVEY OPERATION. **a.** The position area survey includes establishing the orienting line on the ground and determining the horizontal and vertical location of the base piece of each platoon. See FM 6–40 for survey procedure.

b. The orienting line is used to lay the platoons for direction. *An orienting line is merely a line of known direction which is materialized on the ground.* By means of this line and measured or computed base angles, platoons are laid on the base line. The orienting line should be selected so that its distance from any gun is not less than 50 yards. A greater distance would give accuracy. The position commander's aiming circle should be close enough so that the position commander is within voice distance of the platoon. If absolutely necessary, a relay can be established. In some cases it may be necessary to use more than one orienting line in order to lay the platoons.

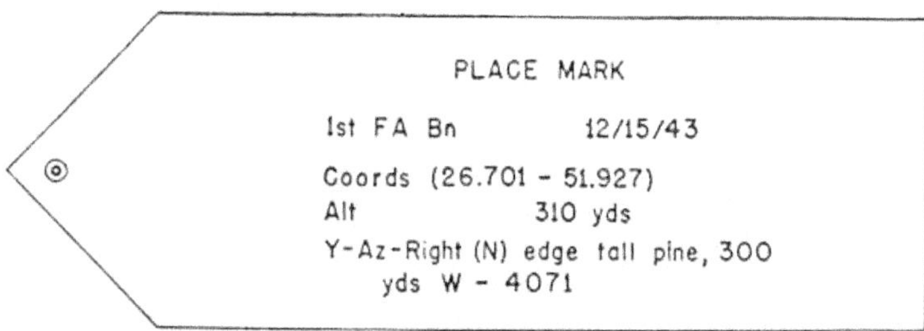

Figure 93. Place mark tag.

c. The horizontal locations of the platoons are determined with reference to the place mark. The angles and distances of the traverse are noted on a sketch.

d. Vertical locations of the platoons are determined as so much higher or lower than the place mark. When the altitude of the place mark has been furnished, the real altitudes of the platoons may be determined and entered on the chart.

195. CONNECTION SURVEY. When the artillery place mark is located at some distance from the company position area, it is necessary for the survey crew to perform a connecting survey to tie in the place mark with the chosen point. (See par. 192b.) The position area survey and the connection survey consist of—

a. Transmitting horizontal and vertical control from the place mark to the platoon positions, and transmitting direction to the orienting line or lines.

b. Computation or measuring of base angles.

196. PLOTTING. When this position area survey has been completed, the positions of the base pieces must be plotted on the firing chart. FM 6–40 lists methods of determining the locations from survey data. Of these methods, the simplest is the use of a large scale plot of the traverse.

197. USE OF MAPS IN SURVEY. A battle map which is known to be accurate always expedites survey. The long traverses for distance often made necessary by a large position area may be abandoned in favor of location of platoon positions by inspection or short traverse from identifiable points near the position. Vertical control, if available from contours, need not be considered in the survey.

198. USE OF AIRPHOTOS IN SURVEY. The same remarks apply when a gridded photomap is used as a firing chart, except that vertical control is carried as in grid sheet survey.

Section III. LAYING THE PLATOON

199. GENERAL. Guns may be laid by base angle or on a Y-azimuth.

a. The procedure for laying on an azimuth is described in paragraph 175.

b. The procedure for laying by base angle is as follows:

(1) The aiming circle is set up on the orienting line.

(2) The announced base angle is set on the azimuth and micrometer scales of the aiming circle.

(3) Using the lower motion, the far end of the orienting line is sighted on. The 0–3200 line of the instrument is now parallel to the base line.

(4) Each gun is laid reciprocally. (See par. 175.)

c. If registration by one gun is feasible prior to the completion of survey, the determination of direction of the orienting line may be avoided. This is done by adjusting one piece (preferably, the piece in the center of the platoon) on the base point. Base deflection is recorded for the center platoon. An orienting line is staked out in any convenient location, and the base angle of the center platoon is *measured.* The base angle of the right platoon is then determined by adding the angle (base piece right platoon—base point—base piece center platoon) to the measured base angle. The base angle of the left platoon is determined by subtracting a similar angle from the measured base angle of the center platoon. The flank platoons are laid on these base angles and base deflection is recorded. FM 6–40 describes the procedure for computation of remaining base angles when a common orienting line is not practicable.

d. When all platoons are registered, an orienting line is not required. Each platoon records the adjusted deflection as base deflection.

e. For greater security against hostile sound or flash ranging, registration may be conducted by a single piece from a surveyed position at a distance from the platoon positions.

200. MEASURING BASE ANGLE. The command received from the OP or fire direction center is MEASURE THE BASE ANGLE. The position commander does not repeat this command. He—

a. Sets up the aiming circle on the orienting line.

b. Commands the base piece AIMING POINT THIS INSTRUMENT, MEASURE THE DEFLECTION. The gunner lays the aiming circle reciprocally.

c. Sights on either end of the orienting line, using the upper motion.

d. Reads the scales of the aiming circle for the measured base angle.

Section IV. REGISTRATION

201. GENERAL. The object of registration is to place the center of impact of one gun on a base point or check point so that corrections determined on the basis of the registration may be used for subsequent firing.

202. ADJUSTMENT. a. Fire is conducted by a single piece. The elevation or gunner's quadrant is used. The normal method of fire is by single rounds.

Each burst is sensed for range. Because of the relatively small burst of 76-mm projectiles, it is often necessary to fire an initial "spotting round" of smoke.

b. The object of adjustment is to obtain a trial elevation; that is, an elevation giving a target hit, or an elevation for the center of a one "c" bracket. To facilitate splitting the bracket, "c" may be taken to the nearest even number.

c. Registration may be accomplished by precision or forward observation methods. The procedure for precision registration is described in FM 6–40. When time is short and ammunition is limited, a platoon may be considered registered on a base point or check point when a target hit has been obtained, or a 25-yard deflection bracket and a 100-yard range bracket has been split by forward observation methods.

d. Artillery and air observers may conduct the registration if they have suitable observation and communications.

e. A report of the results of registration must be made to the fire direction center as early as possible so that computation of corrections can begin.

f. If targets outside the transfer limits of the base point are to be fired upon, registration should be conducted on a check point located near them. If this is not practicable, accuracy must not be expected in firing on such targets. (For transfer limits, see FM 6–40.)

g. Registration will provide accurate corrections only as long as weather conditions remain reasonably constant. It may be advisable, therefore, to check registrations shortly before opening fire or to apply the effects of weather changes to initial registration data.

203. TRANSFER OF FIRE. **a.** A transfer of fire is the application of corrections determined from registration to the firing chart data of a target.

b. The principle underlying the use of corrections determined by registration is that firing on any point of known map location furnishes adjusted data which are different from the basic data determined from the firing chart. These differences have many causes. From these differences, corrections are determined to be applied to map data for other targets within transfer limits.

c. Site as announced in basic data is not corrected.

d. Corrections are computed and applied by the FDC. The procedure is described in FM 6–40.

Section V. FIRE-DIRECTION CENTER

204. GENERAL. **a.** The firing chart is maintained by the fire-direction center. It is the function of the FDC to—

(1) Plot the platoon positions.

(2) Plot the locations of concentrations furnished by the reinforced artillery.

(3) Compute basic firing data.

(4) Compute corrections based on registration.

(5) Apply these corrections to the data determined for the required concentrations.

b. When fully constituted, the FDC is composed of the following members:

(1) Gunnery officer.

(2) Horizontal control operator (HCO).

(3) Vertical control operator (VCO).

(4) One computer for each of the three platoons.

205. OPERATION. The method of operation of a field artillery battalion fire-direction center is discussed in FM 6–40. The organization of the company FDC depends upon personnel available. There may be only sufficient personnel for a horizontal control operator and three computers. When personnel are limited, their duties in the preparation and correction of prearranged or unobserved fires are as follows:

a. Horizontal control operator. (1) Plots the concentrations to be fired on the firing chart. (See par. 206.)

(2) Announces to all computers—

Transfer point.

Deflection correction.

Ammunition.

Number of concentration to be fired.

Unit to fire.

Number of volleys.

Time to fire.

Example:

TRANSFER FROM BASE POINT.

CORRECTION ALL PLATOONS, LEFT EIGHT.

CONCENTRATION ONE ZERO FOUR.

COMPANY.
THREE ROUNDS PER GUN PER MINUTE.
H MINUS 5 TO H.

(3) Announces to each computer—
Range measured from firing chart.
Deflection shift from firing chart.
Angle of site.

Example:

SUGAR
FIVE SIX SIX ZERO.
LEFT ONE SIX FOUR.
SITE PLUS SIX.

FOX
FIVE SEVEN TWO ZERO.
LEFT ONE SEVEN FOUR.
SITE PLUS FIVE.

TARE
FIVE FOUR SIX ZERO.
LEFT ONE EIGHT SEVEN.
SITE PLUS EIGHT.

b. Computer. (1) Repeats to the HCO each element of the data that concern only his particular platoon.

(2) Enters on the data sheet the commands and data received from the HCO.

(3) Applies the corrections to the range and deflection.

(4) Fills in the command section of the data sheet.

(5) Sends the commands to the firing platoon by phone or by data sheet.

(6) If his platoon has been used for registration, the computer—

(*a*) Announces to the other computers the graphical firing table setting determined by the registration.

(*b*) Announces to the HCO the deflection correction as determined from the registration.

206. DESIGNATION OF TARGET LOCATIONS. a. Target data are given by the reinforced artillery. They may be transmitted by—

(1) Overlay, showing locations and altitudes of each target with sufficient marginal notes to clarify matters such as time of firing and ammunition allotment.

(2) Listing of coordinates, altitudes, and other pertinent information of each target.

b. The latter method sometimes results in errors because of misunderstanding of coordinates or errors in plotting but has the advantage that data may be transmitted by telephone.

c. On an overlay, the center of the concentration is given. When coordinates of a target are given, they denote the location of the center of the target. When coordinates of a barrage are given, they usually denote the center point of the barrage line.

d. The procedure for plotting concentrations on the firing chart is described in FM 6–40.

207. DATA SHEETS. a. The FDC computes *corrected*

data for all concentrations. The information may be entered on a data sheet. (See FM 6–40.)

b. Copies of the data sheet are sent to each platoon.

208. FIRING ON ORDER. Prearranged fires on call may be delivered on the order FIRE CONCENTRATION NUMBER____. If no ammunition allotment has been prearranged for this mission, the amount should be prescribed.

209. AMMUNITION. During execution of this secondary mission, units such as tanks and tank destroyers do not use organic loads of ammunition. However, if additional ammunition is available at the firing position, gun crews may fire from ready racks and immediately restow the racks from the ground. It is more convenient to pass ammunition into the turret from outside the gun carriage than to withdraw the rounds from stowage positions other than ready racks. Crews should habitually refill ready racks during every lull in firing so that they may be ready to perform their primary mission.

PART FIVE

AUXILIARY WEAPONS

CHAPTER 20

MECHANIZED FLAME THROWERS

210. GENERAL. Mechanized flame throwers as *auxiliary weapons* may be mounted in the bow machine gun port, in the assistant driver's periscope, or in the gunner's telescopic sight port. As a *primary weapon,* it may be mounted in a dummy gun which replaces the tank cannon. (See current publications as listed in FM 21–6 for detailed technical data, procedure for installation, changing, operation, maintenance, tactical employment, crew drill, and other information.)

211. METHODS OF MOUNTING. a. Bow replacement flame gun (fig. 94). The gun unit of the flame thrower is interchangeable with the bow machine gun and is mounted in place of this weapon when in use. Mounted in this manner, the traverse of the flame gun is limited to the traverse of the bow machine gun (ball) mount. This restricts the operation of the flame thrower to a small angle of traverse and necessitates moving the tank to engage targets outside the limits of traverse. When the flame gun is used instead of the bow machine gun, the fire power of the tank is diminished by one weapon. However, the flame gun does not permanently replace the bow machine gun. The two can be interchanged in approximately 1 minute by a trained gunner.

Figure 94. Bow replacement flame gun.

b. Periscope mounted flame gun. Mounting the flame gun alongside the assistant driver's periscope permits a greater angle of traverse and also retains all of the basic armament of the tank. This type of mounting gives more accuracy than the bow replacement type, since the periscope may be used for sighting.

c. Telescopic sight mount. The flame gun when mounted in the coaxial sight port is coaxially mounted with the cannon. It has greater range than bow mounted flame guns and has the additional advantage of 6,400 mils traverse. The flame thrower and the cannon are sighted by using the periscopic sight.

d. Cannon mount. In the flame throwing tank, the flame gun replaces the cannon. The flame gun is mounted in a dummy cannon. The cannon mount has the advantages of greater range, 6,400 mil traverse, and increased fuel capacity. The gunner's periscope is used for sighting.

212. TECHNIQUE OF FIRE. a. General. (1) Flame throwers are primarily short range direct fire weapons. Fire may be placed on reverse slopes using observed or unobserved methods. Enemy troops on reverse slopes and behind walls, or similar shelter, are attacked at maximum effective range so that the flame will drop behind the crest of the hill or top of the wall.

(2) A cross wind disperses the stream of flame and may make ignition unreliable. An experienced operator can adjust for most wind conditions.

(3) Unthickened fuel produces a dispersed short-range flame. When the fuel is thickened, the stream of flame becomes more "rodlike," its range increases, and the persistence of the burning fuel on the point of impact becomes greater. Thickened fuel produces more intense heat on the target and much less smoke. In general, thickened fuels have a 50 percent greater range than unthickened fuels.

b. Control. The flame gunner normally attacks targets upon the order of the crew commander. In an emergency, he may bring fire to bear as rapidly as possible without command from the crew commander. All members of the crew are alert for flame thrower targets and, when one is sighted, bring it to the attention of the crew commander over the interphone. The crew commander decides how to attack targets outside the field of fire of the flame gun.

c. Conservation of fuel. Flame thrower operators must conserve fuel rigidly. The types of targets encountered will determine the time length and number of bursts that will be necessary for destruction or neutralization. The extremely short duration of fire makes it imperative that a minimum amount of fuel be used on each target.

d. Headwinds, crosswinds, and tailwinds. The direction and velocity of the wind must be taken into consideration when operating the flame thrower. It should not be fired into a strong headwind or crosswind, especially when unthickened fuel is being used. Firing into a strong wind should be avoided since burning fuel may fall short of the target or even blow back toward the firing vehicle. Unthickened fuels cannot be fired into a headwind of more than five to eight miles per hour. When firing at or near maximum ranges, crosswinds deflect, break up, and disperse flame, reducing both range and accuracy. (See fig. 95.) Tailwinds increase the range of the flame.

e. Effective ranges. The ranges of the various guns are dependent upon whether thickened or unthickened fuel is used and on the direction and velocity of wind. An effort should be made to get as close as possible to the target to increase ac-

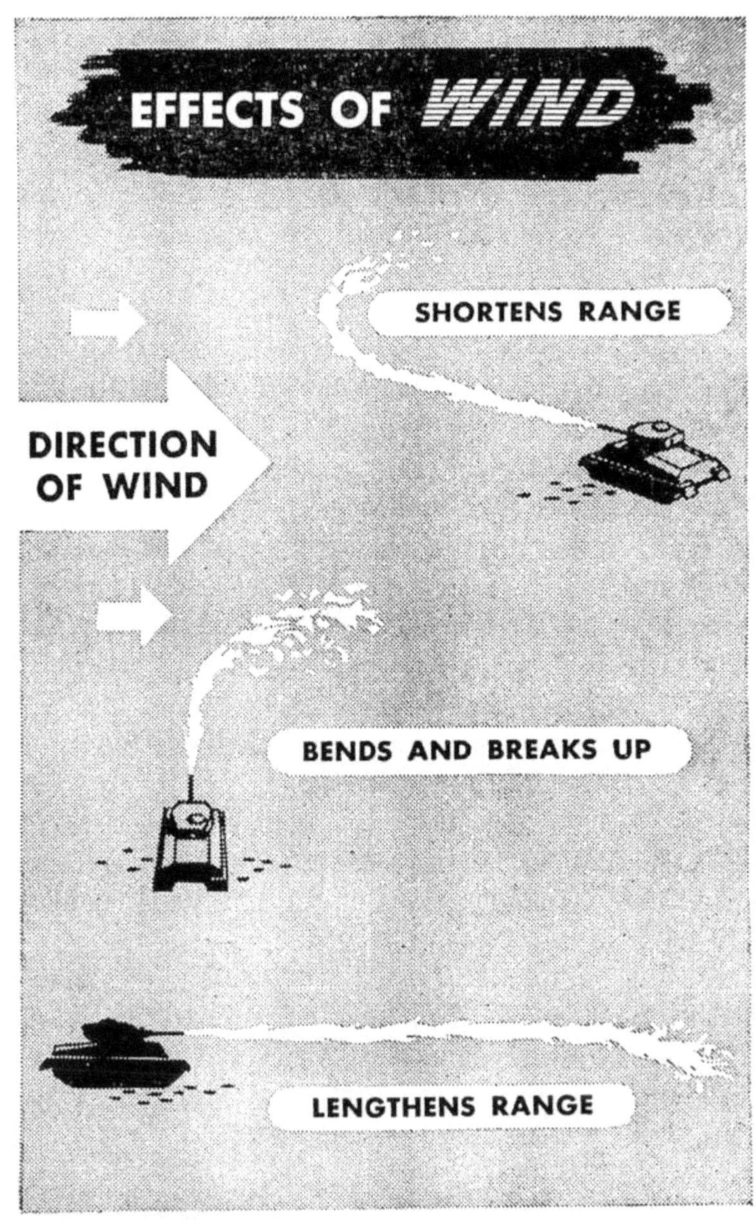

Figure 95. Wind effects on flame.

curacy and to conserve fuel. As a general guide, bow-mounted flame throwers can fire effectively with a favorable tailwind up to 70 yards with thickened fuel, and up to 30 yards with unthickened fuel. The turret-mounted flame throwers can be fired up to 150 yards with a tailwind.

f. Operation. The bow-mounted flame throwers are operated by the bow gunner. The turret-mounted flame throwers are operated by the gunner.

g. Adjusting fire. The course of the attack and the nature of the target will determine the method of adjusting fire on the target. Since the only methods of sighting provided for the weapon are the telescope and the periscope, the gun should be discharged in short bursts of one second or less. This will enable the flame gunner to range in on his target with a minimum of fuel wastage and will allow observation of the burst effect, which is impossible with sustained fire. The first burst should be fired to secure an "over" to prevent obscuration of the target.

213. TACTICAL EMPLOYMENT. For tactical employment of mechanized flame throwers, see FM 17–30.

214. TRAINING. The best training for flame thrower gunners is actual firing on different type targets on varied terrain and under varying weather conditions.

CHAPTER 21

ARMORED VEHICLE ROCKET LAUNCHERS

215. 7.2-INCH MULTIPLE ROCKET LAUNCHER, M17.

a. Characteristics. The 7.2-inch Multiple Rocket Launcher, M17 is designed to be mounted on the Medium Tank, M4, M4A1, M4A3, M4A4, and M4A6. (See fig. 96.) It has a capacity of 20 rockets which can be fired electrically in either single or automatic fire. The launcher is controlled in azimuth and elevation with the same controls as those used for the cannon. It can be jettisoned by means of hydraulic controls operated from within the turret of the tank. All rockets may be fired within 10 seconds, using ripple fire. For details of installation, description, functioning, operation, ammunition, care and maintenance, see TM 9–396.

b. Ammunition. (1) *Classification.* Rockets for use with the 7.2-inch Launcher, M17 are issued in the form of unassembled complete rounds. The high explosive rocket is intended for use against reinforced concrete constructions and fortifications. The chemical rocket contains a chemical charge with a burster to open the shell and distribute the contents at the target. The practice rocket shell contains inert filler and dummy fuze, but is otherwise similar to the service round.

(2) *HE, T37 w/Fuze, Rocket BD, MK146.* This rocket has an initial muzzle velocity of 160 feet per second and a maximum range of approxi-

mately 200 yards. It contains approximately 32 pounds of filler.

c. Special precaution. The following special safety precautions are to be noted:

(1) Fire at temperature above 10° F. or below 120° F.

Figure 96. 7.2-inch Multiple Rocket Launcher M17, mounted on medium tank.

(2) Before loading, remove the safety plug from the firing mechanism box. This safety plug must be kept in the personal possession of the gunner. It should be inserted in the firing mechanism box only when ready to fire. As soon as firing ceases, the safety plug is removed by the gunner.

(3) All hatches are closed prior to firing.

(4) To avoid the rearward blast of the rocket, personnel and combustible material must be cleared to a distance of 60 yards.

(5) The launcher will not be jettisoned onto the rear deck. This can be accomplished by traversing the turret until the launcher is at right angles to the longer axis of the tank.

(6) Duds and those rounds from which the propellant charges have been fired should be handled only by trained Ordnance personnel.

CHAPTER 22

ARMORED VEHICLE MACHINE GUNS

216. GENERAL. Tanks are usually equipped with caliber .30 bow and coaxially mounted machine guns. They also have caliber .30 or caliber .50 machine guns mounted on the top of the turret for antiaircraft or ground fire. Tank destroyers are usually equipped with only a caliber .50 machine gun mounted on the turret top for firing at targets in the air and on the ground. For description, disassembly, assembly, care, cleaning, functioning, stoppages, immediate action, and machine gun gunnery, see FM 23–55. For similar information on the caliber .50 Browning machine gun, see FM 23–65. Pertinent field manuals cover the technique of placing in action guns mounted in armored vehicles, the use of the guns in vehicle evacuation, and in dismounted action.

217. BOW MOUNTED MACHINE GUN. a. General. The bow machine gun protrudes through the hull and is fired by the assistant driver (bow gunner). For mounting, see FM 23–55. This gun has an elevation of approximately 360 mils and a depression of about 180 mils. It can be traversed approximately 270 mils in either direction.

b. Technique of fire. In firing the gun, the bow gunner should initially depress the gun by raising the pistol grip and observing the strike or tracer. He should be trained to know the approximate position of the gun for the desired range. As soon as strikes or tracers are observed, he elevates or depresses the muzzle until the strike or tracer stream is moved to the target. Terrain conditions affect observation of strikes. In dry, dusty terrain, strikes are observed easily. When the ground is wet or covered with grass and other vegetation, strikes are extremely difficult to observe. When strikes cannot be observed, reliance must be placed upon the observation of tracers. Accordingly, it is necessary to fire longer bursts while ranging-in.

c. Fire orders. Fire orders to the bow gunner are normally issued over the interphone. For sequence and example, see section II, chapter 8.

d. Targets. Both offensive and defensive missions can be fired with the bow machine gun. It is generally employed against short range targets and is seldom fired from a tank at ranges in excess of 600 yards. Offensive targets include personnel and unarmored and lightly armored vehicles. In the final assault, the bow gun should always be fired to maintain fire superiority upon the objective unless friendly personnel are endangered. The gun can also be used in reconnaissance by fire of likely hostile installations from either a stationary or a moving tank. Used in conjunction with the coaxial machine gun, it is excellent for repelling attacks by personnel against the tank or against other tanks which the tank may be covering. The ground mount permits it to be removed from the tank and used in a perimeter defense. It also may serve as a crew weapon if the tank has been abandoned

218. COAXIAL MACHINE GUN. a. General. The coaxial machine gun protrudes through the turret and mantlet and is fired by the gunner. Firing is normally accomplished with the firing solenoid. It may be fired manually if the solenoid becomes inoperative. Since this gun is in a fixed mount, it is capable of delivering more effective fire upon point targets than is the bow gun.

b. Adjustment. The process of boresighting and sight adjustment of the cannon automatically gives a correct sight adjustment for the coaxial machine gun if the machine gun has been mounted properly.

(1) Machine guns in a fixed mount having no adjustments are considered to be correctly adjusted.

(2) Machine guns in adjustable mounts are adjusted for elevation as follows:

(*a*) The normal sight adjustment is made.

(*b*) The M1 quadrant is placed on the breechring of the cannon.

(*c*) The bubble is leveled with the elevating handwheel.

(*d*) The set screw located at the rear of the machine gun trunnions is tightened with the fingers. The locknut on the set screw is tightened with a wrench. The machine gun can then be removed from the cradle for cleaning without readjusting the set screw each time.

(*e*) Without disturbing the level of the cannon, the quadrant is transferred to the cover of the coaxial machine gun.

(*f*) By means of the elevation adjusting nuts

located on the rear trunnion bracket, the machine gun is elevated or depressed until the bubble is level.

(*g*) The elevation adjusting nuts are firmly locked against the trunnion bracket or the adjusting screw is secured by tightening the set screws. The preceding adjustment mechanically boresights the cannon and the machine gun.

(*h*) After final adjustment and tightening, the machine gun is rechecked with the M1 quadrant to make sure that it has not moved during the tightening of the nuts.

(*i*) For greatest accuracy, the final adjustment should be checked by firing at a known range.

(3) Some coaxial machine gun mounts have adjustments for deflection and elevation. The deflection for the machine gun can be determined by "firing in" or by use of a testing target. An alternate method is to boresight the machine gun for deflection at the same time that the cannon and its sight are boresighted.

c. Technique of fire. The coaxial machine gun can be fired effectively at ranges up to the tracer burn-out point (800–1,000 yards). It can be fired in excess of this range if the strike can be observed. The use of this gun in moving tank fire is covered in paragraph 149. In firing from a stationary tank at a stationary target, the gunner ranges in on his target by observing the point of strike and the tracer stream. Firing the machine gun from a stationary or moving tank at a moving target is covered in paragraph 158.

d. Fire orders. For sequence of commands and an example of fire orders, see section II, chapter 8.

e. Targets. The principal targets of the machine gun are personnel and unarmored or lightly armored vehicles. It is employed constantly in reconnaissance by fire. Special note should be made of its use to defend the tank or other tanks against attacks by "tank-destroying personnel teams" or "tank hunters." This machine gun, together with the bow gun, is fired in the final assault to maintain fire superiority on the objective unless friendly troops are endangered. It is capable of opening such fire at greater ranges than is the bow gun and can place more effective fire upon the objective.

219. ANTIAIRCRAFT .30 OR .50 CALIBER MACHINE GUN. This weapon is mounted outside the tank and is fired by the tank commander. It is fired as a free weapon and considerable skill is required in directing fire. The weight of the tank commander against the grip will reduce the vibration. Changes in direction are made by moving the body. Sudden shifts throw the gunner off balance and result in erratic shooting. For ground targets, the ranging-in method of observing the tracer stream or point of strike is employed. For aerial targets, adjustment is by observation of the tracer stream. For training and technique of fire against aerial targets, see FM 23–55 and 23–65.

APPENDIX I

SAFETY OFFICER

Section I. INDIRECT FIRE

1. PURPOSE. **a.** During training, a safety officer should be at each firing position. The safety officer has no other duties. He oversees the safe firing of the guns.

b. It is the responsibility of the safety officer to see that the safety precautions prescribed in AR 750–10 are complied with.

2. GENERAL DUTIES. The general duties of a safety officer are to see that—

a. No gun fires outside the lateral safety limits prescribed by the safety card, firing memorandum, or overlay.

b. No gun fires above the maximum elevation prescribed by the safety card, firing memorandum, or overlay.

c. No gun fires below the minimum elevation *computed* by the position commander, or prescribed by the safety card, whichever is greater.

d. The announced powder charge is used.

e. No gun is fired unless the range is clear.

f. No personnel is in those portions of the danger area visible from the position except as authorized in AR 750–10.

g. All personnel at the position comply with safety regulations.

3. SAFETY CARD. The safety officer obtains the following data from the safety card (fig. 97), firing memorandum, or overlay.

a. The date and time that firing is authorized for the position. This is important because safety limits often vary from day to day, and the use of

SAFETY CARD

DATE: *3-6-44* FIRING POINT: *#6* AREA: *HAYS SCHOOL*

REFERENCE POINT: *BLUE MOUNTAIN* y-az: *2750*

LEFT LIMIT: *1110 m̸ LEFT OF REF PT*

RIGHT LIMIT: *300 m̸ LEFT OF REF PT*

MINIMUM RANGE: *3000* yards ______ mils

MAXIMUM RANGE: *7000* yards ______ mils

SPECIAL INSTRUCTIONS: *1110 m̸ LEFT OF REF PT TO 640 m̸ LEFT OF REF PT. MAXIMUM RANGE 6500 YARDS 75 MM TANK GUN, M3, SHELL, M48 (SUPER) 0830-1700*

Card prepared by *EJO'B* Checked by *WDP*

Figure 97. Safety card.

a specific firing point is authorized only for the dates and times given on the safety card, firing memorandum, or overlay.

b. The area in which the designated firing point is located.

c. The specific firing point from which firing is authorized.

d. Where possible, a reference point is usually given. The principal use of the reference point is to tie in the safety limits with the terrain. The approximate Y-azimuth of the reference point is given.

e. The right and left lateral limits are given either as Y-azimuths or in relation to the reference point. The lateral limits mark the limits of deflection shifts. No gun may fire to the right of the right limit or to the left of the left limit.

f. The maximum and minimum ranges are given in yards. The safety officer computes the maximum and minimum elevations from this information.

g. Any special instructions such as different maximum or minimum ranges for specific sectors are given last.

h. The safety card, firing memorandum, or overlay is signed or initialed by the officer who prepares it and the officer who checks it. This is the safety officer's assurance that the card is correct.

4. SAFETY DIAGRAM. The safety officer takes the data given on the safety card and constructs a diagram of the safety limits which he will use. The safety officer puts all the necessary information on the safety diagram, combining the information on the safety card with the data obtained at the firing position. The elements shown on the safety diagram (fig. 98) are as follows:

a. The lateral limits are shown as rays drawn out from the firing position with the Y-azimuth marked on them.

b. The maximum and minimum ranges are

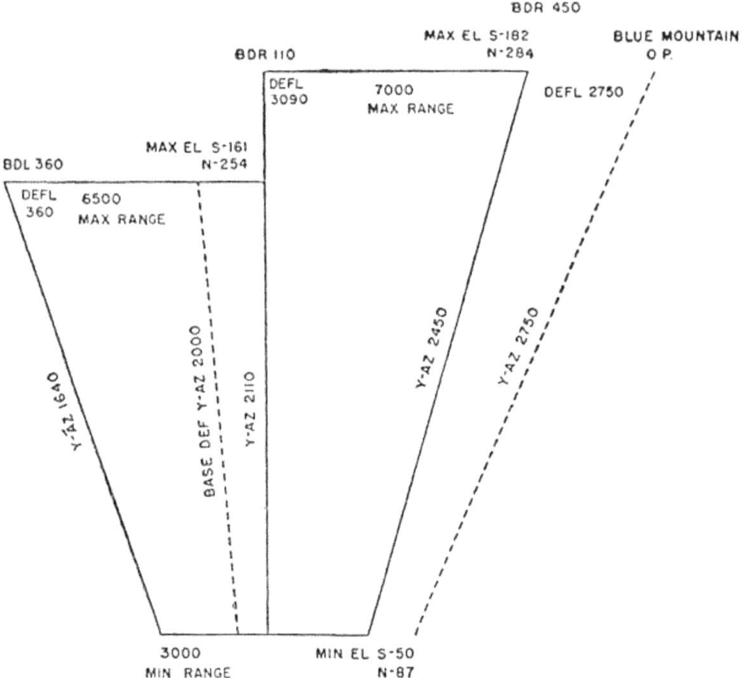

Figure 98. Safety diagram.

shown as arcs with the maximum and minimum range written on them.

c. Any azimuth marking a change on the maximum or minimum range is shown by a ray exactly like those rays marking the lateral limits, and the Y-azimuth is marked on it.

d. The maximum elevation obtained from the firing tables for the range prescribed on the safety card is entered at the top of the diagram. The maximum elevation is obtained for all authorized types of ammunition.

e. The minimum elevation is entered at the bottom of the diagram. The minimum elevation is either the elevation for the minimum range given on the safety card, firing memorandum, or overlay, or the minimum elevation computed by the position commander, *whichever is the greater*. The minimum elevation is computed for all authorized types of ammunition.

f. The minimum and maximum elevations may be changed by applying corrections determined from registration. The officer in charge of firing is responsible that the safety officer receives the correct data.

g. The reference point is indicated by a ray similar to the lateral limit rays. The nature of the reference point and the Y-azimuth are entered on this line.

h. As soon as the platoon has been laid and the safety officer has verified the laying, he computes the deflections of the lateral limits and of any intermediate points at which the maximum or minimum elevations change. For example:

(1) The platoon has been laid on compass 2000, and all gunners have set their azimuth indicators at zero after being laid.

(2) If the left safety limit is Y-azimuth 1640, a shift of left 360 mils lays the gun on the left limit.

(3) If the right limit is Y-azimuth 2450, a shift of right 450 mils lays the gun on the right limit.

(4) In like manner if the Y-azimuth of the intermediate ray of the safety diagram is 2110, a shift of right 110 mils lays the gun on the intermediate ray.

i. If base deflection has been recorded, the safety officer determines the shifts from base deflection to all lateral limits. The base deflection shifts of critical points may be determined in a like manner.

5. SAFETY STAKES. **a.** The safety officer marks the lateral limits of each gun by means of safety stakes. These stakes are marked in a distinctive color and shape for each gun and are labeled R, L, and I, for right, left, and intermediate.

b. When the platoon has been laid, the safety officer announces RIGHT SAFETY STAKE DEFLECTION____, LEFT SAFETY STAKE DEFLECTION____, INTERMEDIATE SAFETY STAKE DEFLECTION____. The crew commander records these commands and has the gunner traverse to the announced deflection to line in the stake for that deflection. When the crew commander has lined in all stakes, he reports SAFETY STAKES READY TO BE CHECKED, NUMBER____. To prevent errors in laying being carried over to safety limits the safety officer checks the proper placement of the stakes for each gun by an independent means, such as aiming circle and azimuth, deflection from reference point.

c. After the safety officer has checked the safety stakes of the guns, he places chalk marks on the turret and on the hull at the right and left safety limits to provide a quick check on whether a gun is within its safety limits.

d. During training, when fire is being conducted over exposed personnel, the safety officer should have minimum elevation stops applied to the guns.

6. PROCEDURE DURING FIRING. a. General. After the safety officer has completed his preparation for firing, he notifies the position commander that he is ready. His principal duty now is to watch the platoon to insure that all firing is conducted in accordance with safety regulations.

b. Guns outside safety limits. (1) If a gun is laid outside one of the lateral limits, the safety officer reports to the officer conducting fire UNSAFE TO FIRE, ____MILS OUTSIDE LEFT (RIGHT) SAFETY LIMIT.

(2) If a command for elevation below the minimum or above the maximum elevation is given, the safety officer reports to the officer conducting fire UNSAFE TO FIRE, MINIMUM (MAXIMUM) ELEVATION____.

c. Special situations. (1) If at any time the range flag (or the range light at night), is lowered, the safety officer orders CEASE FIRING until the flag (light) is raised.

(2) If at any time the range flag is obscured, the safety officer orders firing suspended until communication is established with the range officer. Where there is no range flag, communication is established before firing commences.

(3) If at any time unauthorized vehicles or personnel are observed in the impact area, the safety officer orders CEASE FIRING until the range is clear.

d. General procedure. (1) The safety officer verifies the initial laying of the platoon. He may check the position commander as the platoon is being laid, or he may measure the compass upon which the platoon has been laid.

(2) The safety officer checks the computation of the minimum elevation by the position commander to avoid any chance of error.

(3) When the safety officer is sure that all guns are laid within the safety limits, he announces SAFE TO FIRE.

(4) If the officer conducting fire assumes responsibility for firing a round outside the safety limits, the safety officer records the time of the occurrence.

e. Equipment for safety officer. (1) An accurately declinated aiming circle.

(2) Safety card.

(3) Copy of current range memorandum to include firing points, time of firing, and responsible officer.

(4) Copy of range regulations with which safety officer should be familiar.

(5) Firing table.

Section II. ESTABLISHING SAFETY LIMITS

7. GENERAL. Safety limits for any range or firing point may be given from a point or from the ends of a line.

8. SAFETY LIMITS FROM A POINT. When safety limits are given from a point, the Y-azimuths of the right and left limits determine the limits of the sector of fire. Generally, guns must be placed within 100 yards of the point. The sector of fire for each gun is determined directly from the Y-azimuths of the safety limits.

9. SAFETY LIMITS FROM A LINE. **a.** When safety limits are given from a line, the sector of fire cannot be determined directly from the azimuths of the safety limits.

(1) The left safety limit is established for guns firing from the left end of the line.

(2) Similarly, the right safety limit is established for guns firing from the right end of the line.

b. For this reason, a gun firing from the *right* end of the line cannot use the given azimuth of the *left* safety limit. The same is true for the right safety limit of guns on the left end of the line.

c. There are two methods of insuring that all guns fire within the safety limits.

(1) Channelizing the sectors of fire.

(2) Measuring the respective safety limits from the opposite ends of the line.

d. When fire is channelized, the full limits of the designated impact area are utilized. However, the sector of fire for each gun is extremely small. The procedure for channelizing fire is as follows:

(1) Using the designated safety limits, the sector of fire is divided into equal parts.

(2) Each gun is assigned the sector to its direct front.

(3) Fire for each gun is restricted to its assigned sector.

e. Measuring the azimuths from the opposite ends of the line has one advantage. It permits a wider sector of fire for any individual gun. It also permits crossfire. However, it does not permit full use of the designated impact area. Procedure is as follows:

(1) The *right* safety limit is measured from the *left* end of the line.

(2) The *left* safety limit is measured from the *right* end of the line.

(3) These safety limits are defined on the terrain.

10. SAFETY LIMITS FOR INDIRECT FIRE. In indirect fire, all guns are laid from a common point and on a common azimuth. Guns are traversed to the right and left safety limits. Safety stakes are put out. (See par. 5, app. I.)

APPENDIX II

BALLISTICS

1. GENERAL. Ballistics is the science or study of the motion of projectiles. Interior ballistics deals with the motion of the projectile in the tube; exterior ballistics with the motion of the projectile after leaving the tube. Expanding powder gases, sealed in the tube by the rotating band, propel the projectile. The rifling, cutting into the rotating band, gives spin to the projectile. Conditions in the tube which change these operations affect the flight of the projectile. If the projectile or the tube is dirty, the projectile may be centered improperly; reduction in range and inaccuracy in direction may result. Injuries to the rotating band, by permitting the escape of gases, may have similar effects.

2. DISPERSION ERRORS. a. Dispersion errors are errors that are inherent in the dispersion pattern. (See fig. 37.) They are the result of variations of certain elements from round to round, even though conditions are as nearly identical as possible. Dispersion errors should not be confused with *mistakes or constant errors.* These are not inherent in the dispersion pattern. *Mistakes* can be eliminated by care and training.

b. For practical purposes, the dispersion error of a shot is the distance from that shot to the center of impact. A dispersion error may be resolved into its range and deflection components.

3. JUMP. **a.** In most guns when the projectile leaves the muzzle, it does not follow the exact line of elevation or deflection at which the gun was laid at the moment of firing. This displacement of the gun tube is referred to as "jump."

b. Jump consists of two elements, a vertical displacement and a horizontal displacement. The vertical displacement is usually the larger. It is also more important since it affects the firing table elevations. The reticles of some coaxial and periscopic telescopes are designed to compensate for jump.

c. There are many causes for variations in the amount of jump. Some of these are differences in—

(1) Design of the gun carriage.

(2) Position of the trunnions.

(3) Ammunition.

(4) Lengths and construction of gun tubes.

d. Differences in the amount of jump require extreme care in the selection of the proper firing table. For example: The firing table elevations for the 76-mm Gun Motor Carriage, M18, are not the same as they are for the same gun mounted on the M4 tank.

e. Because of variations in jump, care must be taken to insure that the proper sight is used. This can be accomplished by comparing the elevations shown on the aiming data chart (figs. 18 and 19) with the firing table elevations for the particular gun and gun carriage.

4. RATES OF FIRE. Because of differences in muzzle velocity and construction, cannon are capable of varying rates of fire. Although muzzle velocity has virtually no effect on maximum rates of fire per minute for a limited period of time, it does influence rate of fire over extended periods. The greater the muzzle velocity the lower the prolonged rate of fire.

☆ U. S. GOVERNMENT PRINTING OFFICE : 1945—675383

ISBN#978-1-937684-56-3
www.PeriscopeFilm.com

www.ingramcontent.com/pod-product-compliance
Lightning Source LLC
LaVergne TN
LVHW010054110826
845155LV00028B/327

* 9 7 8 1 9 3 7 6 8 4 5 6 3 *